Advanced Smart Computing Technologies in Cybersecurity and Forensics

Advanced Smart Computing Technologies in Cybersecurity and Forensics

Edited by

Keshav Kaushik
Shubham Tayal
Akashdeep Bhardwaj
Manoj Kumar

CRC Press
Taylor & Francis Group
Boca Raton London New York

CRC Press is an imprint of the
Taylor & Francis Group, an **informa** business

First edition published 2022
by CRC Press
6000 Broken Sound Parkway NW, Suite 300, Boca Raton, FL 33487-2742

and by CRC Press
2 Park Square, Milton Park, Abingdon, Oxon, OX14 4RN

CRC Press is an imprint of Taylor & Francis Group, LLC

Library of Congress Cataloging-in-Publication Data
Names: Kaushik, Keshav, editor. | Tayal, Shubham, editor. | Bhardwaj,
    Akashdeep, 1971- editor. | Kumar, Manoj (Professor of computer science), editor.
Title: Advanced smart computing technologies in cybersecurity and forensics / edited by Keshav
    Kaushik, Shubham Tayal, Akashdeep Bhardwaj, and Manoj Kumar.
Description: First edition. | Boca Raton : CRC Press, 2022. | Includes bibliographical references and
    index. | Summary: "This book addresses the topics related to artificial intelligence, internet of
    things, blockchain technology, and machine learning and brings together researchers, developers,
    practitioners, and users interested in cybersecurity and forensics"-- Provided by publisher.
Identifiers: LCCN 2021036513 (print) | LCCN 2021036514 (ebook) | ISBN 9780367686505 (hardback)
    | ISBN 9780367690137 (paperback) | ISBN 9781003140023 (ebook)
Subjects: LCSH: Computer security. | Computer crimes--Investigation.
Classification: LCC QA76.9.A25 A354 2022 (print) | LCC QA76.9.A25 (ebook)
    | DDC 005.8--dc23/eng/20211007
LC record available at https://lccn.loc.gov/2021036513
LC ebook record available at https://lccn.loc.gov/2021036514

ISBN: 978-0-367-68650-5 (hbk)
ISBN: 978-0-367-69013-7 (pbk)
ISBN: 978-1-003-14002-3 (ebk)

DOI: 10.1201/9781003140023

Typeset in Times
by SPi Technologies India Pvt Ltd (Straive)

To Those Who Inspired This Book but Will Never Read It!

Contents

Preface

Advanced Smart Computing Technologies in Cybersecurity and Forensics (ASCTCSF) by CRC Press, Taylor & Francis Group, USA is a brainchild of Keshav Kaushik, Dr. Shubham Tayal, Dr. Akashdeep Bhardwaj, and Dr. Manoj Kumar. Advanced practices of cybersecurity and forensics help solve real-life problems and provide solutions to secure user data and help our daily lives, to make the world a better, more secure place. Here we share our ideas in form of a book.

Big thanks to all our co-authors, who are experts in their own domains, for sharing their experience and knowledge. This book is an attempt to compile their ideas in the form of chapters to share with the world. This book provides insights into Machine Learning, Artificial Intelligence, and real-time new age domain to supplement Cyber Security Operations, Governance, Compliance, Policies, Risk Management, and VA/PT. The book also looks at Cybersecurity for Big Data, IoT and Blockchain, and Threat Modeling. The book will be helpful for security professionals, academicians, scientists, advanced-level students, penetration testers, and researchers working in the field of cybersecurity and forensics.

Acknowledgments

I have to start by thanking first my parents and then my wife for their immense support and giving me the time and space to work on this book and my research. My wife looked after handling the house, groceries, and kids, to reading early drafts, giving me advice, which was positively useful. A big shout out to CRC Press, Taylor & Francis Group, and especially the journals managing editor, who kept up with my constant spam emails and diligently guided and supported me. I am honored to be a part of this book's journey with all my coauthors, so thank you for letting me serve, for being a part of CRC Press, and for putting up each time with my irritating endless emails and help myself and my co-authors turn their ideas into stories.

Editor Biographies

Mr. Keshav Kaushik is an Assistant Professor in the Department of Systemics, School of Computer Science at the University of Petroleum and Energy Studies, Dehradun, India. He is pursuing a Ph.D. in Cybersecurity and Forensics. He is an experienced educator with over six years of teaching and research experience in Cybersecurity, Digital Forensics, the Internet of Things, and Blockchain Technology. Mr. Kaushik received his B. Tech degree in Computer Science and Engineering from the University Institute of Engineering and Technology, Maharshi Dayanand University, Rohtak. In addition, his M. Tech degree in Information Technology is from YMCA University of Science and Technology, Faridabad, Haryana. He has qualified in GATE (2012 and 2016). He has published several research papers in International Journals and has presented at reputed International Conferences. He is a Certified Ethical Hacker (CEH) v11, CQI, and IRCA Certified ISO/IEC 27001:2013 Lead Auditor, and a Quick Heal Academy-certified Cyber Security Professional (QCSP). He has acted as a keynote speaker and delivered professional talks on various national and international platforms.

Dr. Shubham Tayal is an Assistant Professor in the Department of Electronics and Communication Engineering at SR University, Warangal, India. He has more than 6 years of academic/research experience teaching at the UG and PG level. He has received his Ph.D. in Microelectronics and VLSI Design from National Institute of Technology, Kurukshetra, M.Tech (VLSI Design) from YMCA University of Science and Technology, Faridabad and B.Tech (Electronics and Communication Engineering) from MDU, Rohtak. He has qualified GATE (2011, 2013, 2014) and UGC-NET (2017). He has published more than 25 research papers in various international journals and conferences of repute, with many papers currently under review. He is on the editorial and reviewer panel of many SCI/SCOPUS indexed international journals and conferences. Currently, he is editor of four books from CRC Press (Taylor & Francis Group, USA). He acted as keynote speaker and delivered professional talks on various forums. He is a member of various professional bodies such as IEEE, IRED, etc. He is on the advisory panel of many international conferences. He is a recipient of Green ThinkerZ International Distinguished Young Researcher Award 2020. His research interests include simulation and modelling of Multi-gate semiconductor devices, Device-Circuit co-design in digital/analog domain, machine learning and IOT.

Dr. Akashdeep Bhardwaj is Professor (Cybersecurity and Digital Forensics) at the University of Petroleum & Energy Studies (UPES), Dehradun, India. Dr. Akashdeep is an eminent IT industry and academic expert with over 26 years of experience in Cybersecurity, Digital Forensics and IT Management Operations. In his current role, Dr. Akashdeep mentors graduate, masters, and doctoral students from national and international universities apart from leading Cybersecurity projects. Akash has

xiv
Editor Biographies

published several research papers, books, chapters, and patents. Akash has worked as a Technology Leader and Head at several multinational organizations.

Dr. Manoj Kumar obtained his Ph.D. (Jan. 2019) in Computer Science from The Northcap University, Gurugram. He did his B. Tech in computer science from Kurukshetra University. He obtained M.Sc. (Information Security and Forensics) degree from ITB, Dublin in and M. Tech from ITM University. Mr. Kumar has 9.5+ years of experience in research and academics. He published over 30 publications in reputed journals and conferences. He published two books and five patents with his team. He has acted as a keynote speaker and delivered professional talks on various forums and conferences. Presently, Mr. Kumar is Assistant Professor (SG), (SoCS) at the University of Petroleum and Energy Studies, Dehradun. Before that, he was engaged as Assistant Professor in the computer science department of AMITY University, Noida. He is a member of various professional bodies and has served on the review board for many reputed journals like *IEEE Transaction on Multimedia* (IEEE Signal Processing Society 2019), *Multidimensional Systems and Signal Processing* (Springer), *Applied Computing and Informatics* (Elsevier), *Multimedia Systems* (Springer), *Kubernetes* (Emerald, 2019), *Journal of Digital Imaging (JDIM)*, (Springer), *IET Intelligent Transport Systems*, and *Journal of Experimental & Theoretical Artificial Intelligence* (Taylor & Francis). He is an Editorial Board Member for *The International Arab Journal of Information Technology* (IAJIT, SCIE Indexed), Jordon, and *Journal of Computer Science Research*, Singapore. He was recognized as a Quarterly Franklin Member (QFM) by London Journal Press in March 2019 and a Bentham Ambassador (India) on behalf of Bentham Science Publisher.

Contributors

C. Aarti
Amity University
Noida, India

P. S. Apirajitha
Department of Information Science and
Technology
CEG Campus, Anna University
Chennai, India

Rangel Arthur
Faculty of Technology (FT)
State University of Campinas
(UNICAMP)
São Paulo, Brazil

Akashdeep Bhardwaj
University of Petroleum and Energy
Studies
Dehradun, India

Billel Bengherbia
University of Medea
Médéa, Algeria

Omar Benzineb
University of Blida
Blida, Algeria

Brahim Belhorma
High School of Signals (HSS)
Koléa, Algeria

Mounir Bouhedda
University of Medea
Médéa, Algeria

Rahul Chand
The University of South Pacific
Fiji

Prachi Chauhan
College of Technology
G.B. Pant University of Agriculture and
Technology
Pantnagar, India

J. Charu
Amity University
Noida, India

Mukesh Chouhan
Government Polytechnic College
Sheopur, India

Bhagwati Garg
Manager-Union Bank of India
Gwalior, India

Rakesh Garg
Amity University
Noida, India

L. Ancy Geoferla
RMK Engineering College
Chennai, India

Sam Goundar
RMIT University
Vietnam

Bhavya Gururani
Amity University
Noida, India

Yuzo Iano
School of Electrical Engineering and
Computing (FEEC)
State University of Campinas
(UNICAMP)
São Paulo, Brazil

G. Maria Jones
Saveetha Engineering College
Chennai, India

Priyanka Joshi
Indian Institute of Technology Indore
Indore, India

Vinod Mahor
IPS College of Technology and
Management
Gwalior, India

Hardwari Lal Mandoria
College of Technology
G.B. Pant University of Agriculture and
Technology
Pantnagar, India

Bodhisatwa Mazumdar
Indian Institute of Technology Indore
Indore, India

Ana Carolina Borges Monteiro
School of Electrical Engineering and
Computing (FEEC)
State University of Campinas
(UNICAMP)
Campinas, São Paulo, Brazil

Alok Negi
National Institute of Technology
Srinagar, India

Nehinbe Joshua Ojo
Federal University
Oye, Nigeria

Kiran Pachlasiya
NRI institute of Science and
Technology
Bhopal, India

Reinaldo Padilha França
School of Electrical Engineering and
Computing (FEEC)
State University of Campinas
(UNICAMP)
São Paulo, Brazil

Durgesh Pandey
Amity University
Noida, India

S. Punitha
Department of Computer Science and
Engineering
Karunya Institute of Technology and
Sciences
Coimbatore, India

Supriya Raheja
Amity University
Noida, India

Romil Rawat
Shri Vaishnav Vidyapeeth
Vishwavidyalaya
Indore, India

Manveer Singh
The University of South Pacific
Fiji

S. Sunil
Amity University
Noida, India

Shrikant Telang
Shri Vaishnav Vidyapeeth
Vishwavidyalaya Indore
India

Stephan Thompson
Department of Computer Science and
Engineering
Faculty of Engineering and Technology,
M. S. Ramaiah University of Applied
Sciences
Bangalore, Karnataka, India

S. Godfrey Winster
SRM Institute of Science and
Technology
Chennai, India

1 Detection of Cross-Site Scripting and Phishing Website Vulnerabilities Using Machine Learning

J. Charu, S. Sunil, and C. Aarti

Amity University, Noida, India

CONTENTS

1.1   INTRODUCTION

The usage of the web has advanced rapidly because of the accessibility in various domains, for example, web-based banking, amusement, instruction, programming downloading, and long-range interpersonal communication. As needs be, a colossal volume of data is downloaded and transferred continually to the web. This gives open doors for crooks to hack significant individual or monetary data, for example, usernames, passwords, account numbers, and public protection numbers. This is known as a web phishing assault, which is considered one of the serious issues in web security (Jian Mao et al. 2017, Mohith Gowda et al. 2020).

Figure 1.1 illustrates the steps of the phishing process. In a web phishing assault, phishing sites are made by the assailant, which are like the authentic

DOI: 10.1201/9781003140023-1

1

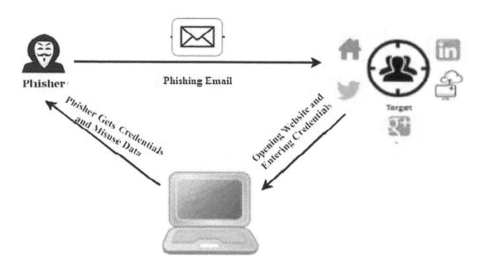

FIGURE 1.1   Steps of Web phishing process.

sites to misdirect web clients to get their delicate monetary and individual data. The phishing assault is at first performed through clicking a connection embedded inside advertisements or messages or emails. Casualties get an email containing a connection to refresh or approve their data. On the off chance that this connection is clicked by the objective casualties, the web program will divert them to a phishing site that seems like the first site. The assailants would then be able to take the significant data of the web clients since they are approached to enter the touchy data on the phishing site. At last, the assailants can do monetary burglary in the wake of phishing happens (Nick Williams and Li 2017, Xin Mei Choo et al. 2016).

Because of the certainty of phishing sites focusing on online organizations, banks, web clients, and government, it is fundamental to forestall web phishing assaults in the beginning phases. In any case, identification of a phishing site is a difficult undertaking because of the numerous imaginative techniques utilized by phishing assailants to mislead web clients.

The achievement of phishing site location methods predominantly relies upon perceiving phishing sites precisely and inside an adequate timescale (Zou Futai et al. 2016). Although researchers have proposed many phishing prevention techniques based on conventional black and white listing or heuristic-based phishing detection techniques yet phishing is a growing challenge as every second new websites are launched by hackers, attackers or criminals. Nonetheless, these strategies are not proficient enough, and new intelligent strategies need to be devised to prevent phishing and protect every user over the internet. Our work aims at designing a machine learning approach, which works well in the predication of XSS done in phishing websites.

1.2 RELATED WORK

Jain Mao et al. (2017) proposed a web page phishing detection system that uses page component similarity value to analyze URL tokens. The prototype is based on a comparison of the CSS style of target pages and involves a phishing alarm in the Google Chrome browser. The authors experimented with real-world phishing samples to prove their results.

Zou Futai et al. (2016) worked on a real network traffic dataset of a large ISP. Through the data cleaning phase, they extracted eight dataset attributes named Reference URL (REF), User Agent (UA), User node number (AD), Visiting URL (URL), User SRC IP (SRC-IP) access time (TS), access server IP (DST-IP) and User cookie (cookie). Their technique was Graph Mining with Belief Propagation (BP) algorithm which constructs an undirected graph of user nodes and URL nodes. The algorithm iterates to correct node reputation to finally check phishing websites.

An ACT-R cognitive behavior architecture model was designed by Nick Williams and Li (2017). The system was prepared in Lisp to mimic human brain behavior with respect to phishing website detection. The judgment that the webpage was real or phished depends on the presence of the HTTPS padlock webpage validity indicator. The authors carried out three experiments to test their proposed system is effective in detecting genuine and fake web pages. The characteristics of HTTPS padlock like color, size, location, and verifiability were exploited, and results had shown that the phishing use case maps well with ACT-R.

Akashdeep Bhardwaj et al. (2016a) reviewed Cryptographic algorithms especially Symmetric and Asymmetric algorithms used in cloud-based applications. They used file size, encryption computation time, and encoding computation time parameters for comparison purposes. RSA and AES encryption algorithms were found to work best for ensuring security in cloud-based applications. MD5, the hashing algorithm, provided the fastest results during the comparison of encoding techniques.

Akashdeep Bhardwaj et al. (2016b) studied and found that cloud environments are most vulnerable to denial-of-service attacks. The authors investigated many existing DDoS attacks and propounded a simpler distributed denial-of-service (DDoS) taxonomy which enables the classification of DDoS attacks. Further, the set of parameters for the DDoS mitigation solution were proposed.

Xin Mei Choo et al. (2016) studied various features used for classifying phishing websites. The authors prepared a feature-based phishing detection technique which extracts some previous existing features and new aggregate features from web pages. They called their feature set a sensitive feature set. Further, they proposed a support vector machine to improve the classification process, and results showed that their system accuracy in detecting fake and genuine websites was 95.33.

Ashna Antony (2020) had proposed a Random Forest machine learning technique for detection and blocking of any old and new phishing URLs. UCI repository dataset was used for the testing proposed model. The author also deployed the model over the cloud, which works with the Google Chrome extension.

In the Real-Time Client-Side Phishing Prevention system, Giovanni Armano et al. (2016) had placed an add-on facility in the browser to detect whether a website is legitimate or not. This system extracts information from the websites visited by the user and warns them with a message for any kind of phishing website. The authors had designed a prototype to detect the target of a phishing website. If the webpage is detected as phished website page, then any interactions with the page will be prevented through a darkened black semi-transparent layer over the webpage.

Mohith Gowda et al. (2020) proposed a browser named "Embedded Phishing Detection Browser (EPDB)," which uses 30 important properties of URL for predicting fake websites. The authors started with datasets of phishing websites like phish tanks and millers miles. Further, 30 properties of URL were taken through the rule of extraction framework. The 30 extracted properties were input into a random forest classifier model, which delivered an accuracy of 99.36% for website classification as real or fake. The model had got 0% false-negative rate. Authors had compared their system with other classification models like Logistic Regression and Support Vector Machine. The browser also warns the user through a pop-up message.

Parasmani M. Rathod et al. (2017) discussed various phishing detection approaches along with their advantages and disadvantages. According to the author's study, there is no single anti-phishing technique which can detect all types of phishing websites. Moreover, maximum techniques work only with English language-based websites. Hence, there is a requirement to support different language websites, and also, the existing techniques need to be modified from time to time to handle cybercrimes.

Ping Yi et al. (2018) used real traffic flow from ISP and introduced two types of features (original and interaction features) in web phishing. Their detection model was based on a deep learning technique – Deep Belief Networks. DBNs were implemented for effective feature extraction, and finally, classification was done as positive and negative for genuine and fake websites. The model was evaluated through four criteria: accuracy, true positive rate (TPR), false-positive rate (FPR), and positive predicative value (PPV).

Hawanna et al. (2016) designed a novel approach for checking a particular HTTP URL using Google's updated blacklist check, Alexa Ranking, Google search engine results, and multiple URL features. The proposed algorithm works well with only HTTP URLs and generates warning messages whenever a website is classified as phishing.

In Jun Hu et al. (2016) a phishing website was detected by referring to log details available on a legitimate website server. The logs are used for extracting details about resources whenever a person becomes a victim by opening a phishing site. Moreover, during the detection of phishing sites, these log files get updated by filling entries of phishing sources by the genuine website servers. This helps in tracking any new phishing websites.

Off-the-Hook application was modified by **S. Suganya et al.** to address various drawbacks like compromising user privacy, difficulty in handling dynamic websites, etc., found in existing phishing detection approaches. Authors have worked on existing Off-the-Hook add-on tools to provide better accuracy and exploit brand independence and language independence features. They proposed a Genetic Algorithm to classify the URLs as genuine versus fake. The system provides resilience to dynamic phish and on-demand services to users.

Peter Likarish et al. (2008) introduced B-APT, a Bayesian Anti-Phishing Toolbar based on Bayesian filters. They studied that Bayesian filters had proved the best results in spam filtering and hence worked on designing a new phishing detector toolbar. It was a Mozilla Firefox extension which consists of a B-APT engine and a user interface. Author results showed that the toolbar detected 100% phishing websites in comparison with IE and Firefox.

XSS attacks are 40% among all web vulnerabilities, according to Cisco 2018 Annual Security Report. German Rodriguez et al. (2020) designed artificial intelligence-based detection and mitigation technique for XSS attacks. The authors gave a detailed description of tools used to mitigate XSS attacks by analyzing 67 documents. They discussed stealing cookies through cross-scripting attacks and different methods of defense against CSS attacks.

1.3   IMPLEMENTATION

XSS (**Source: Acunetix**) is a common web attack done by injecting malicious code in client-side scripts. The attacker targets the embedded scripts in a vulnerable website to take control over the victim's application and sensitive data. When the victim visits the infected webpage or website, the malicious code gets executed through the client browser and causes the web attack. A web application or web page is XSS vulnerable if it asks for user inputs which are not validated before actual submission. The input data is parsed at the victim's browser and gets executed along with malicious code. XSS attacks are possible in HTML, CSS, VBScript, JavaScript, ActiveX and Flash. The most prevalent XSS vulnerable scripting language is JavaScript, as it is used in creating almost every webpage.

1.4   PHISHING WEBSITES DETECTION

1.4.1   PHISHING WEBSITES

The quantity of phishing assaults has been filling significantly as of late and is considered as one of the most hazardous current web wrongdoings, which may lead people to lose trust in the web-based business. Subsequently, it has a gigantic negative impact on online trade, showcasing endeavors, associations' wages, connections, clients, and general business activities.

To take the visitor identities and certifications, the phisher normally builds up a fake imitation of the first site, which is comparable in appearance to the first site. Consequently, the phisher sends an email to many victims to criminally perform deceitful monetary exchanges in the interest of the web clients. Some web users get trapped, and they become a victim of phishers attack.

1.4.2   PHISHING WEBSITES DETECTION TECHNIQUES

It is imperative to identify the phishing sites right on time, to caution the clients against sending their delicate data through these phony sites. The adequacy and exactness of phishing sites recognition strategies are pivotal for the achievement of phishing identification.

A few traditional methods for distinguishing phishing sites have been proposed in writing to adapt to the web phishing issues. Be that as it may, the choice with respect to the phishing sites in these strategies was anticipated loosely. Hence, two mainstream approaches are utilized to recognize the phishing sites:

- *Blacklist and whitelist-based approach*: This methodology does rejection or acceptance of a web source while visiting it and categorize it as either a phishing or authentic site individually. Here blacklist refers to a list of distrustful or malicious web objects which ought to be stopped from running on networks. Whitelist deals with trustful web objects and blocks anything which is not present in its maintained list. The primary downside of the blacklist and whitelist-based methodology is that it cannot recognize the recently made phishing sites from real sites. (Source: Consolidated Technologies, Inc.)
- *Intelligent heuristics-based approach*: This methodology gathers important elements or information of sites which can be used to assess the website as phishing or genuine site. A dataset is prepared with the help of numerous web sources. This dataset is further used by researchers to design effective phishing detection systems which are capable of recognizing new phishing sites too (Eduardo Luzeiro Feitosa et al. 2019).

1.5   IMPLEMENTATION FLOWCHART (Figure 1.2)

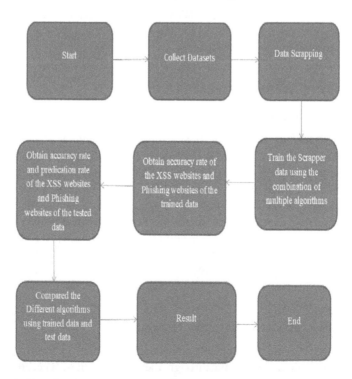

FIGURE 1.2   Proposed methodology for detection of web phishing.

1.5.1 Dataset

The datasets are obtained from XSSed.com, where the data are collected from the year 2009–2013. The dataset contains details of XSS attacks through 200 websites. The number of features extracted is 22, which consists of script-based features, DOM objects, URL-based features, etc.

1.5.2 Classifiers

Approaches of detection of XSS and phishing web site using machine learning approaches:

- Random Forest Classifier,
- MLP Classifier,
- ELM Classifier,
- Naïve Bayes,
- Support Vector Machine,
- K Neighbors Classifier.

1.6 RESULT AND DISCUSSION

Experiments are carried out over the XSS dataset using six machine learning algorithms, i.e., Random Forest Classifier, MLP Classifier, Gen ELM Classifier, Naïve Bayes, Support Vector Machine, and K Neighbours Classifier with feature vector comprising of 22 features.

Naïve Bayes classifiers are actually a collection of algorithms derived through the Bayes theorem and are widely used in the field of statistics, analytics, and probability theory. These classifiers have proved best in solving email spam filtering, sentiment analysis, and document classification. Their best part is they help in evaluating important parameters in less amount of time with even small training datasets (Gyan Kamal and Manna (2018), Source: Naïve Bayes Classifiers). Emmanuel Gbenga Dada et al. (2019) survey showed that Naïve Bayes has achieved high accuracy (99.99+%) in spam classification and also has low FPR.

Support Vector Machines are one of the most powerful supervised learning models that are effortlessly trained during data analysis and data classification (Emmanuel Gbenga Dada et al. 2019). SVM has proved better than many classification techniques used for email spam identification. However, the literature also shows that SVM suffers from computational complexity during the processing of data.

Extreme Learning Machine (ELM) Classifiers have properties like self-adaptive, low computational times, and high accuracy (Chapala Maharana et al. 2017). It has many varieties like Twin Extreme Learning Machine (TELM), Pruned Extreme Learning Machine (PELM), Class-Specific Cost Regulation ELM (CCRELM), and Multilayer Extreme Learning Machine (ML-ELM). Out of these, TELM has proved comparable results with SVM and TSVM.

Random Forest algorithms are also used in data analysis and classification domains. These algorithms result in less classification error, good f-score values, and

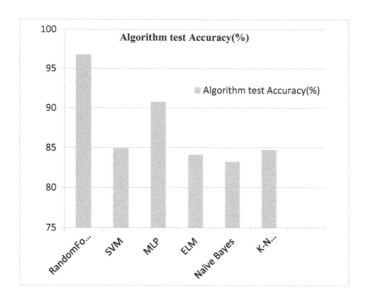

FIGURE 1.3   Graph showing accuracy comparison of six machine learning algorithms.

fast execution speed. (Emmanuel Gbenga Dada et al. 2019). This work results that the Random Forest algorithm gives the highest accuracy test as compared to the other algorithms (Figure 1.3).

1.7   CONCLUSION AND FUTURE WORK

Research on phishing and XSS has been a big challenge these days. The objective of this research work is checking a given URL is a phishing website or not. This paper's emphasis is on XSS attacks and phishing attacks on web pages, important features of the web document, and URL used in different machine learning and other techniques. The work demonstrates the XSS Scan technique used to locate different features which help to identify XSS infected webpage.

The comparison of the chosen six algorithms, such as Random Forest Classifier, MLP Classifier, ELM Classifier, Naïve Bayes, Support Vector Machine, and K Neighbors Classifier, shows that the Random Forest Classifier is giving the highest test accuracy, i.e., 96.7%. This work can be further extended by finding more features from the selected dataset. The experiment can also be conducted by using deep learning algorithms to enhance user privacy and support the safe use of dynamic web pages. The experimental results have proved that Multilayer perceptron and Random Forest algorithm are giving a high accuracy rate when compared to Naïve Bayes classifier, SVM, ELM classifier, K-N Neighbour classifier.

REFERENCES

Antony, Ashna. (2020). Detecting Phishing Websites using Data Mining. *International Research Journal of Engineering and Technology (IRJET)*, Vol. 7 (6): 1762–1764.

Bhardwaj, Akashdeep, Gampa Subrahmanyam, Vinay Avasthi, Hanumat Sastry. (2016a). Security algorithms for cloud computing. *Procedia Computer Science*, Vol. 85: 535–542.

Maharana, Chapala, Ch. Sanjeev Kumar Dash, Bijan Bihari Mishra. (2017). Extreme learning machine classifier: a topical state-of-the-art survey. *International Journal of Engineering Science Invention*, 48–59.

Dada, Emmanuel Gbenga, Joseph Stephen Bassi, Haruna Chiroma, Shafi'i Muhammad Abdulhamid, Adebayo Olusola Adetunmbi, Opeyemi Emmanuel Ajibuwa. (2019). Machine learning for email spam filtering: review, approaches and open research problems. *Heliyon* (Elsevier), Vol. 5 (1): 23.

Rodriguez, German, Jenny Torres, Pamela Flores, Diego Benavides. (2020). Cross-Site Scripting (XSS) attacks and mitigation: a survey. *Computer Networks*, 166, ISSN 1389-1286.

Kamal, Gyan, Monotosh Manna. (2018). Detection of phishing websites using naïve Bayes algorithms. *International Journal of Recent Research and Review*, Vol. 11 (4): 34–38.

Feitosa, Eduardo Luzeiro, Carlo Marcelo Revoredo da Silva, Vinicius Cardoso Garcia. (2019). Heuristic-based strategy for phishing prediction: a survey of URL-based approach. *Computers & Security*, 88: 1–20.

Mao, Jian, Wenqian Tian, Pei Li, Tao Wei, Zhenkai Liang. (2017). Phishing-alarm: robust and efficient phishing detection via page component similarity. *IEEE Access*, Vol. 5: 17020–17030.

Marchal, S., G. Armano, T. Gröndahl, K. Saari, N. Singh, N. Asokan. (2017). Off-the-hook: An efficient and usable client-side phishing prevention application. *IEEE Transactions on Computers*, Vol. 66 (10): 1717–1733.

Gowda, Mohith, Heggur Ramesh, M.V. Adithya, Prasad S. Gunesh, S. Vinay. (2020). Development of anti-phishing browser based on random forest and rule of extraction framework. *Cybersecurity*, Vol. 3: 1–14. https://cybersecurity.springeropen.com/track/pdf/10.1186/s42400-020-00059-1.pdf (Accessed 17 September 2021).

Rathod, Parasmani M., Arnab Gajbhiye, Rahul Atri, Anita Sachin Mahajan. (2017). A survey of phishing website detection and prevention techniques. *International Journal of Innovative Research in Science, Engineering and Technology*, Vol. 6 (9): 19069–19073.

Yi, Ping, Yuxiang Guan, Futai Zou, Yao Yao, Wei Wang, Ting Zhu. (2018). Web phishing detection using a deep learning framework. *Wireless Communications and Mobile Computing*, Vol. 2018: 1–9.

Choo, Xin Mei, Kang Leng Chiew, Dayang Hanani Abang Ibrahim, Nadianatra Musa, San Nah Sze, Wei King Tiong. (2016). Feature-based phishing detection technique. *Journal of Theoretical and Applied Information Technology*, Vol. 91 (1): 101–106.

CONFERENCES

Bhardwaj, Akashdeep, G.V.B. Subrahmanyam, Vinay Avasthi, Hanumat Sastry, Sam Goundar. (2016b). *DDoS attacks, new DDoS taxonomy and mitigation solutions – a survey. International Conference on Signal Processing, Communication, Power and Embedded System (SCOPES).* https://www.researchgate.net/publication/318329206_DDoS_Attacks_New_DDoS_Taxonomy_and_Mitigation_Solutions_-_A_Survey (Accessed 17 September 2021).

Armano, Giovanni, Samuel Marchal, Nadarajah Asokan. (2016). *Real-time client-side phishing prevention add-on. IEEE 36th International Conference on Distributed Computing Systems (ICDCS)*, 777–778.

Hu, Jun, Xiangzhu Zhang, Yuchun Ji, Hanbing Yan, Li Ding, Jia Li, Huiming Meng. (2016). *Detecting phishing websites based on the study of the financial industry webserver logs.*

3rd International Conference on Information Science and Control Engineering (ICISCE), Vol. 1: 325–328.

Williams, Nick, Shujun Li. (2017). *Simulating human detection of phishing websites: an investigation into the applicability of ACT-R cognitive behaviour architecture model. Proceedings of 2017 3rd IEEE International Conference on Cybernetics (CYBCONF 2017),* 471–478.

Likarish, Peter, Eunjin Jung, Donald E. Dunbar, Thomas E. Hansen, Juan Pablo Hourcade. (2008). *B-APT: Bayesian anti-phishing toolbar. IEEE International Conference on Communications*: 1745–1749.

Hawanna, Varsharani Ramdas, Vrushali Kulkarni, Rashmi Rane. (2016). *A novel algorithm to detect phishing URLs. International Conference on Automatic Control and Dynamic Optimization Techniques (ICACDOT),* IEEE, 548–552.

Futai, Zou, Gang Yuxiang, Pei Bei, Pan Li, Li Linsen. (2016). *Web phishing detection based on graph mining. 2nd IEEE International Conference on Computer and Communications (ICCC),* 978-1-4673-9027-9: 1061–1066.

ONLINE DOCUMENTS/RESOURCES

Acunetix, Cross-Site Scripting (XSS). https://www.acunetix.com/websitesecurity/cross-site-scripting/. (Accessed 25 February 2020).

Consolidated Technologies, Inc, Blacklisting vs. Whitelisting. (August 2019). https://consoltech.com/blog/blacklisting-vs-whitelisting/. (Accessed 25 February 2020).

How to recognize phishing email messages or links. (March 2011). http://www.microsoft.com/security/online-privacy/phishing-symptoms.aspx. (Accessed 25 February 2020).

Naïve Bayes Classifiers. (2019). https://www.geeksforgeeks.org/naive-bayes-classifiers/#:~:text=Naive%20Bayes%20classifiers%20are%20a,is%20independent%20of%20each%20other.

UCI Machine Learning Repository. (2012). http://archive.ics.uci.edu/ml/. (Accessed 10 April 2020).

XSSED. http://xssed.com/. (Accessed 10 April 2020).

Security Algorithms for cloud computing. (May 2016). Elsevier Procedia Science Direct, 85 (2016): 535–542.

Solutions for DDoS Attacks on Clouds. (July 2016). IEEE International Conference, 978(1), 4673-8203-8/16.

DDoS attacks, new DDoS taxonomy and mitigation solutions – a survey. (October 2016). IEEE Xplore Digital Library, 78-1-5090-4620-1/16.

Reducing the threat surface to minimize the impact of cyber-attacks. (April 2018). Elsevier Network Security Journal (NSW), 2018 (4): 15–19.

Efficient fault tolerance on cloud environments. (June 2018). International Journal of Cloud Applications and Computing (IJCAC), 8 (3): 2.

Framework to define relationship between cyber security and cloud performance. (February 2019). Elsevier Computer Fraud and Security (CFS), 2019 (2): 12–19.

Framework for effective threat hunting. (June 2019). Elsevier Network Security Journal (NSW), 2019 (6): 15–19.

Capturing-the-invisible (CTI): behavior-based attack recognition in IoT-oriented industrial controls systems. (June 2020). IEEE Access, Vol. 8.

Cloud and IoT based smart architecture for desalination water treatment. (February 2021). Environmental Research, Vol. 195 (4).

2 A Review

Security and Privacy Defensive Techniques for Cyber Security Using Deep Neural Networks (DNNs)

Prachi Chauhan and Hardwari Lal Mandoria
G.B. Pant University of Agriculture and Technology,
Pantnagar, India

Alok Negi
National Institute of Technology, Srinagar, India

CONTENTS

2.1 INTRODUCTION

The concept of protecting networks, systems, and applications from digital breaches in cybersecurity. The security measure is often essential because the public, government, business, financial and medical enterprises are storing, processing, and

DOI: 10.1201/9781003140023-2

retaining unprecedented quantities of information on computing devices. A large portion of such data could be confidential information, whether it is the financial documents, personal information, intellectual property, or other forms of information over which unauthorized disclosure or access may have adverse consequences. Presently, the implementation of successful cyber security policies is very difficult since there are numerous devices and attackers becoming much more imaginative [1]. Every day, an innumerable range of threats are produced, the patterns of these are subtly distinct from one another, and it is quite harder to identify such newfound attacks. Typically, cyberattacks are intended to access, manipulate, or damage confidential information, defraud users' money, or disrupt regular business processes. In order to resolve the whole circumstance, the responsibility for cyber security is primarily improved. There is a proliferation of tools such as antivirus, vulnerability scanning, intrusion detection systems (IDSs), and intrusion security devices. However, several attackers are already at a benefit since they just need to identify one flaw in the systems that need security means if the volume of Internet-connected systems [2] grows, the attack space also expands, that leading to a higher probability of attack. A bunch of research work has been done in this area over the last few decades, but these current unknown threats still pose a barrier to researchers. In this regard, researchers have adapted various DL principles [3] for the development of an intelligent system.

DL is a series of machine learning techniques that begin to learn throughout multiple levels, characterized by various abstraction levels, as demonstrated in Figure 2.1. Levels refer to distinctive levels of meanings, whereby higher-level principles are described from lower-level principles, as well as similar lower-level principles can help to describe certain higher-level principles. DL models [4] become more and more precise as they analyze more data, fundamentally learn from past findings to improve their abilities to create correlations and associations. DL also applied to

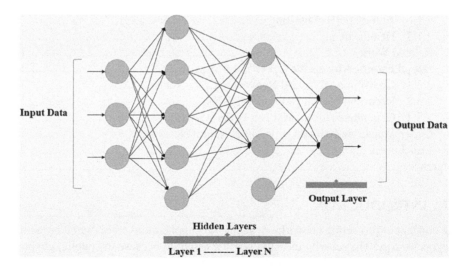

FIGURE 2.1   DL abstraction levels.

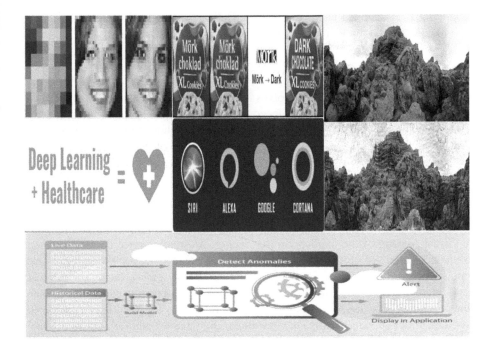

FIGURE 2.2 DL applications field.

almost any area of science and also brought about revolutionary changes. DL improves all aspects of an operation via discovering regular difficulties and also introduces additional domains of analysis, with new excellent progress of DL within the region of comprehensive protection measures. DL has provided outstanding results in various fields.

Here we are going to address the DL applications that dominated the identified markets in 2020 and beyond, which are illustrated in Figure 2.2.

2.1.1 PIXEL RESTORATION

Pixel restoration is the method of upscaling or enhancing the information of an image. Sometimes a low-resolution object image is provided to an input, and then the similar image is scaled up to a higher resolution, and that is the output. Features of the high-resolution results are filled where the specifications are widely undefined. It greatly increases the resolution of images, recognizing prominent features that are only enough to distinguish the characteristic.

2.1.2 DEEP DREAMING

Deep dreaming facilitates the development of hallucinated computer images. Wildly creative visuals are created by a neural network that is actually a series of predictive learning models, driven by deceptively simple algorithms that are modeled after evolutionary cycles. It helps the machine to hallucinate on top of an original picture – thereby producing a reassembled vision of a dream. Hallucination appears to alter based on the type of deep net, including what it has been exposed to.

2.1.3   IMAGE–LANGUAGE TRANSLATIONS

The attractive application of DL involves image–language translation. With the help of the Google Translate app, it has become possible to quickly and automatically translate text-based digital images into a language that choose by you in real-time. Everything that you should do is to set the camera on top of the object, and then the system runs a DL network to read the picture, OCR it (i.e., translate it to text), and convert this into the chosen language.

2.1.4   VIRTUAL ASSISTANTS

A Virtual Assistant is an application that recognizes human voice commands then executes specific tasks for users. Using Natural Language Processing to fit user voice input to perform commands. Digital assistants are simply cloud-based services that need Internet-connected systems or devices to interact with certain capabilities. Each time a request is issued to an assistant, they prefer to have a good user experience based on prior experience using DL algorithms. Digital assistants varying from Alexa to Siri to Google Assistant are the most common DL applications. Every experience with all these assistants provides an opportunity to learn much about your voice and tone, giving you a supplementary human interaction perspective.

2.1.5   FRAUD DETECTION

The financial and banking field, which is troubled by the challenge of detecting fraud in electronic financial transactions, is another DL benefiting domain. Fraudulent financial practices can be identified by looking at on-site and conspicuous signals. Abnormally large transactions, or those that occur at a typical place, clearly deserve extensive verification. Regression and classification DL methods and neural nets are used for the detection of fraud. Although deep learning is often used to illustrate fraud cases involving human negotiation, DL aims to reduce these attempts by scaling up attempts.

2.1.6   AUTOMATIC HANDWRITING

DL has played a crucial role in the processing of automated handwriting. It transforms the user input as text into a handwritten script. The machine automatically records the movement of a pen as well as the characters that users need to remember. Training also promotes the development of new writing forms.

2.1.7   HEALTHCARE

DL is increasingly finding its way into revolutionary technologies with high-value implementations in the real-world medical environment. Researchers are required to increase the performance of automation and smart decision-making in main patient treatment and public health systems. The field of genetic medicine is now so novel that unexpected findings are prevalent, generating a fascinating perfect environment for creative approaches to effective therapy. Electronic Health Record systems (EHR) store patient records, such as demographic data, medical prior history reports, and laboratory results. EHR models increase the rate of accurate diagnosis and also the time needed to create a prognosis with the use of DL algorithms. These technologies use data stored in EHR systems to identify patterns in public health and risks factors and also to make conclusions about patterns they detect.

2.2   RELATED WORK

Fang et al. [5] constructed a new functional encryption method for internal product predicates due to the R-LWE problem. The lattice-based functional encryption method realizes fine-grained encryption and resists quantum attacks. In this development, first, use the configuration algorithm to produce the public parameters and also the primary secret key. Second, evaluate the secret key correlated with the predicate vector v depending on the R-SIS problems using the key generation algorithm. Third, measure the ciphertext aligned with the w-attribute vector depending on the R-LWE problem by using an encryption algorithm. The method reduced the length of the keys and ciphertext compared to the error-based learning scheme.

Bandyopadhyay et al. [6] proposed a whole intrusion detection system approach based on the Optimal Convolution Neural Net (CNN). CNN-IDS has the ability to develop automated intermediate functions but also achieves greater precision in both categories of binary and multi-class. The benefit of the CNN-IDS model has a high detection accuracy with minimum model development time. It operates with just 11 main features rather than 41 features that help to offer higher performance over a very short timeframe. The latest work is compared to the proposed model for detection rates, Recall, F1-Score, Precision, and Support.

Sarker et al. [7] introduced A multi-layer DL system for cyber security simulation. The overall aim was to clarify cyber security predictive analytics and related approaches but also to concentrate on data-driven smart decision-making to protect systems against cyber attacks. Authors have addressed certain areas of research consideration, such as the experimental evaluation of the proposed data-driven system as well as the comparative comparison with some other security measures. For researchers, the multi-layer DL algorithm could be considered as a guide when developing smart cyber security frameworks for organizations.

Rai et al. [2] introduced few modern classifiers, such as XG-Boost, LGBM, that are designed to detect network intrusion, and the performance was evaluated on the basis of an NSL KDD dataset. The authors examined Logistic regression, SGD, XG-Boost, LGBM, DNN as a network-based intrusion detection model. A layered classifier has been formed from all five modeling techniques as a base classification classifier. T Results of experiments showed that the Gradient Boosting Decision Tree (GBDT) configurations, including Light GBM (LGBM) and XG-Boost, and the Stacked Ensemble Classifier, yield the best accuracy rate. The GBDT models have the maximum accuracy of 99 percent for the DOS, Probe, U2R, and R2L sections. The SGD has a minimum accuracy of 90 percent for the probe section. Therefore, this also is inferred that GBDT ensembles are used to build an efficient IDS.

Sharma et al. [8] proposed the GUESS method that is based mostly on GA and secret exchange mechanisms. The mechanism was often used to create a frame series in a quiet way that the association between every two frames was reduced. The frame series describes randomization as a scale of the wide range of video frames. The recommended approach reduces the time needed and increases the efficiency of its encryption algorithm that satisfies the criteria of application areas—for instance, YouTube, video transmission, distribution of highly confidential information by obscuring it in frames. The performance of the experimental system became so compelling that normal data could be encrypted. As a result, the suggested GUESS model requires less time and then becomes more efficient and precise.

Yin et al. [9] developed a DL methodology to predict intrusion via recurrent neural networks (RNN-IDS). In addition, they also reviewed the performance of the method in binary and multi-class classification and the number of features and diverse learning rate effects on the results of the suggested model by comparing it with the performance of the J48 model, ANN, naive Bayes, support vector machine and some other machine learning models throughout the benchmark collection of data. The experimental results showed that RNN-IDS was very useful in predicting a classification model with higher accuracy, and so its efficiency always is preferable to that of conventional classification methods for both binary and multi-class classifications. The RNN model increased the efficiency of intrusion detection and brought a novel research tool for intrusion detection.

Rai et al. [1] explored a summary of cyber threats, how to avoid them, and addressed in-depth the most destructive cyber attacks, some leading cybercriminals of all time, and nations with some of the most cyber-crimes and cyber violations of the 21st century. The authors have introduced how malicious hackers use their technical abilities to commit crimes and remove the idea of technical innovations that must be the key engine of economic development, tracking innovations including certain cloud technology, e-commerce, Big Data analytics, online wallets, AI, automated learning and social networking. There should be some kind of system and protocols to secure humanity from these cyber threats, so there must be a method called cyber protection.

2.3   DEEP LEARNING MODELS FOR CYBER SECURITY

Secure and efficient artificial intelligence is still focusing on cyber-crimes, attacks [10, 11], vulnerabilities. Attacks against DL develop incorrect predictions by inserting false samples and provide gradient-based methods to disrupt the machine. Every corporation prefers to continue their information confidential as well as the rivals do not use it for certain business purposes.

The key concept that DL system is designed to fulfill three key criteria while maintaining privacy:

i.   Cloud server settings should never be exposed to the user.
ii. The stored data throughout the training should not be shared with the cloud server.
iii. The requirement of the user should never be revealed to the cloud server.

The goal of this study is to understand the recent phenomenon of in-depth knowledge of secret data and concerning security issues related to DL in various domains. In addition, we identify various forms of DL potential data security threats alongside different protection methods. DNN demonstrates impressive success in unidentified data. Some common DNN architectures [12] include conv nets, recurrent neural networks (RNNs), deep belief networks, generative adversarial networks, and auto-encoders.

2.3.1 CONVOLUTIONAL NEURAL NETWORKS (CONV NETS)

Conv Nets uses convolution functions in at least one of its layers [3], as compared to conventional neural networks. As shown in Figure 2.3, conv nets use a multi-layer or multi-stage variation consisting of an automated feature extractor as well as a training classifier having necessary components:

The filter bank or kernel's purpose is to identify a specific characteristic at every input position. Therefore the spatial representation of the input from the characteristic identification layer will indeed be transmitted to the output 0 without any modification. As specified by LeCun [4], there seems to be a bank of m_1 filters inside each convolutional layer, as well as the output $Y_i^{(l)}$ of the lth layer consists of $m_1^{(l)}$ feature maps of the size $m_2^{(l)} * m_3^{(l)}$. The map of the Ith function is determined as follows:

$$Y_i^{(l)} = B_i^{(l)} + \sum_{j=1}^{m_1^{(l-1)}} K_{ij}^{(l)} * Y_j^{(l-1)} \tag{2.1}$$

where, $B_i^{(l)}$ indicates the trainable bias parameters matrix, $K_{ij}^{(l)}$ is the filter with dimensions $\left(2h_1^{(l)+1} * 2h_2^{(l)+1}\right)$ that connect the jth feature map of $(l-1)$ layer with ith feature map of (l) layer, and $(*)$ is the 2D discrete convolution operator.

Convolutional Layer is convoluted from the preceding layer at which the input image from its input layer is convoluted to many features of the output image.

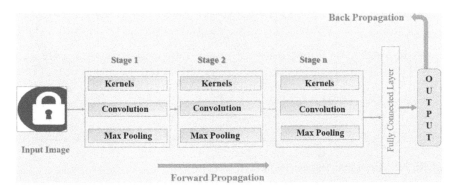

FIGURE 2.3 Convolutional neural network architecture.

Max Pooling Layer will pool all of the highest values of all the components within the preceding layer. Hence, performance after the max pooling layer will be a feature map comprising the most prominent features of a preceding feature map.

Fully Connected Layer also known as output layer and last layer of conv nets. It depends on output aspects of the classification. The inputs to a fully connected layer are indeed the result of the final Pooling or convolutional layer that is flattened and afterward fed into this layer.

2.3.2 RECURRENT NEURAL NETWORKS (RNNs)

It is most often preferred for processing sequential data. As shown in Figure 2.4, the RNNs calculate output value after modifying the currently hidden units, previously hidden units, and presently accessible data means expanding the functionality of the conventional neural nets that only accept fixed-length input data to address variable-length sequence data. RNN [13] handles the data of one unit using the results of the hidden layer as alternate inputs in the following unit at a time. As a consequence, RNNs encounter issues such as speech-language problems, sequence problems as well as gradient loss problems, and short-term memory. The rectified linear unit is being used to address these problems.

Input: $x(t)$ is often used as input to the neural network at time step t.

Hidden state: $h(t)$ depicts a hidden state at time step t and functions as a network "memory." $h(t)$ is determined on the basis of the current input and the hidden state of the prior time step:

$$h(t) = f\left(U\,x(t) + W\,h(t-1)\right) \tag{2.2}$$

The function f is taken to be a non-linear transformation such as ReLU.

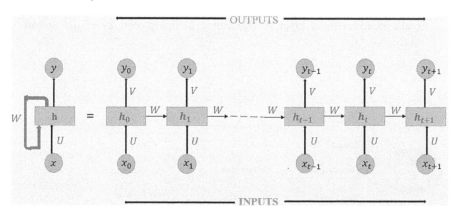

FIGURE 2.4   Recurrent neural network architecture.

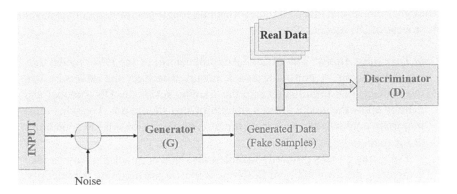

FIGURE 2.5   GANs architecture.

Weights: The RNN inputs hidden connections weighted by the weight vector U, hidden-to-hidden persistent connections weighted by the weight vector W, and hidden-to-output connections weighted by the weight vector V, and these weighted parameters (U, V, W) are transmitted over time.

Output: $y(t)$ represents the output performance of the network at time step t.

2.3.3   GENERATIVE ADVERSARIAL NETWORKS (GANs)

GAN comprises essentially of two neural network configurations, as shown in Figure 2.5, which remain competitive against one another in a zero-sum game to outsmart one another. In it, one network functions as a generator (G) and the other as a discriminator (D) [14]. The generator collects input data in the framework and generates output data of the similar characteristic as the actual data while discriminator is being used within the architecture to decide whether or not the generator data is real? Once training is completed, the generator is able to produce new data that cannot be separated from real data. GAN has been chosen for several fields [15], such as image processing, voice recognition, and domain adaptations.

2.4   CYBER ATTACKS AND THREATS WITH DEEP NEURAL NETWORK

DNNs are becoming prominent for video surveillance, pattern recognition, smart agriculture [16], object detection, and image analysis like medical image analysis [17], IoT-Oriented Industrial Control Systems [18, 19]. However, a new study suggests that DL networks can be breached by deliberately planned critical attacks with slight unnoticeable distortions. It prompts privacy concerns regarding the deployment of such applications in technological settings.

The cyber threats can indeed be categories into four types throughout the development process of DL models:

Data Poisoning Attack: This attack takes full control of the DNN model during the training. A poisoning attack occurs whenever the adversary can insert inaccurate information into the training set of the DNN model and thereby allow the model to learn something that should not. The other most important outcome of a poisoning threat is that variations in the boundaries of the model.

Poisoning attacks primarily include two kinds of attackers: Anyone that is targeting the availability of DNNs as well as some targeting the integrity.

Backdoor Attack: Backdoor is a malware category that mitigates the usual access of the system for authentication procedures. As a consequence, they are granting perpetrators the freedom to remotely enforce system commands and modify malware along with databases and file servers. It means remote access to information inside the frameworks is provided to the attacker. Backdoor threat [20] integrates backdoors into the DNN model in such a manner that both selected sub-task of the attacker, as well as the relatively harmless main task, are learned by the backdoor model. It indicates a backdoor is a form of input from which the developer of the model is really unaware of modification but that the perpetrator can exploit to get the DNN framework to do whatever they want.

Backdoor attacks implant DL models with unknown connections or causes to circumvent logical inferences like classification and enable the algorithm to maliciously function as per the attacker's intention while usually performing in the lack of the target.

Adversarial Examples Attack: Adversarial examples represent inputs toward DNNs algorithms that perhaps the attacker purposely created to allow the algorithm to make mistakes; they are just like visual illusions for computers. DNN algorithms consider inputs as a linear vector. Developing the input in a particular way to get all the incorrect outcomes of the model is indeed an adversarial attack. An adversarial example [21] is also an instance of small, deliberate irregularities that lead a deep neural algorithm to produce a false prediction. The intruder applies a minor disruption which has been measured to allow the target image identified as a separate object image of high accuracy.

Model Stealing Attack: In these threats, adversary through black-box access, but without any previous knowledge of metrics or training data of DNN model, attempts to replicate the features of (namely the "steal") model [22] means that it is used to "steal" models or extract training data representation through black-box scanning. Through many user queries, an intruder attempts to steal a deep neural model. The model would generate predicted labeling with a knowledge level after joining usual queries via prediction APIs.

As illustrated in Figure 2.6, the first two attacker's activities happened during the training process, although the last two threats occurred during the testing process.

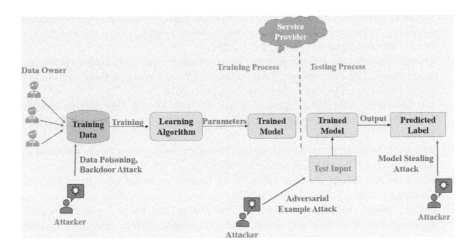

FIGURE 2.6 Attack scenario with DNNs.

2.5 CONCLUSION

Deep neural network-based application systems are widespread and also encounter a range of security attacks across their lifetimes. This chapter presents a systematic and detailed survey of the research work regarding cyber security principles for DL and sheds some attention on DNN security for the complete development cycle of a learning network system. We also summarized the DL approaches to a diverse range of cyber security attacks targeting networks, host systems, software, and data. This chapter will ideally provide detailed recommendations on the design of stable, reliable, and confidential DNN systems. Future studies should acknowledge a cascading association of malicious actions over the existence cycle of an attack.

REFERENCES

1. Rai, Mahima, and Hardwari Mandoria. "A study on cyber crimes cyber criminals and major security breaches." *International Research Journal of Engineering and Technology* 6, no. 7, 1–8, 2019.
2. Rai, Mahima, and Hardwari Lal Mandoria. "*Network intrusion detection: A comparative study using state-of-the-art machine learning methods.*" In *2019 International Conference on Issues and Challenges in Intelligent Computing Techniques (ICICT)*, Vol. 1, pp. 1–5. IEEE, New York, 2019.
3. Goodfellow, Ian, Yoshua Bengio, Aaron Courville, and Yoshua Bengio. *Deep learning*, Vol. 1. Cambridge: MIT Press, 2016.
4. LeCun, Yann, Koray Kavukcuoglu and Clément Farabet. "*Convolutional networks and applications in vision.*" In *Proceedings of 2010 IEEE International Symposium on Circuits and Systems*, pp. 253–256. IEEE, New York, 2010.
5. Fang, Shisen, Shaojun Yang, and Yuexin Zhang. "Inner product encryption from ring learning with errors." *Cybersecurity* 3, no. 1, 1–11, 2020.
6. Bandyopadhyay, Samir, Ratul Chowdhury, Arindam Roy, and Banani Saha. "A step forward to revolutionise intrusion detection system using deep convolution neural network." Preprints, 1–13, 2020.

7. Sarker, Iqbal H., A. S. M. Kayes, Shahriar Badsha, Hamed Alqahtani, Paul Watters, and Alex Ng. "Cybersecurity data science: An overview from machine learning perspective." *Journal of Big Data* 7, no. 1, 1–29, 2020.

8. Sharma, Shikhar, and Krishan Kumar. *"Guess: Genetic uses in video encryption with secret sharing."* In *Proceedings of 2nd International Conference on Computer Vision & Image Processing*, pp. 51–62. Springer, Singapore, 2018.

9. Yin, Chuanlong, Yuefei Zhu, Jinlong Fei, and Xinzheng He. "A deep learning approach for intrusion detection using recurrent neural networks." *IEEE Access* 5, 21954–21961, 2017.

10. Bhardwaj, Akashdeep, and Sam Goundar. "Reducing the threat surface to minimise the impact of cyber-attacks." *Network Security* 2018, no. 4, 15–19, 2018.

11. Bhardwaj, Akashdeep, and Sam Goundar. "A framework for effective threat hunting." *Network Security* 2019, no. 6, 15–19, 2019.

12. Indolia, Sakshi, Anil Kumar Goswami, S. P. Mishra, and Pooja Asopa. "Conceptual understanding of convolutional neural network – A deep learning approach." *Procedia Computer Science* 132, 679–688, 2018.

13. Zhang, Xu-Yao, Fei Yin, Yan Ming Zhang, Cheng-Lin Liu, and Yoshua Bengio. "Drawing and recognizing Chinese characters with recurrent neural network." *IEEE Transactions on Pattern Analysis and Machine Intelligence* 40, no. 4, 849–862, 2018.

14. Goodfellow, Ian J., Jean Pouget-Abadie, Mehdi Mirza, Bing Xu, David Warde-Farley, Sherjil Ozair, Aaron Courville, and Yoshua Bengio. *"Generative adversarial nets."* In *Proceedings of the 27th International Conference on Neural Information Processing Systems*, Vol. 2, pp. 2672–2680. MIT Press, Cambridge, MA, 2014.

15. Yang, Qingsong, Pingkun Yan, Yanbo Zhang, Hengyong Yu, Yongyi Shi, Xuanqin Mou, Mannudeep K. Kalra, Yi Zhang, Ling Sun, and Ge Wang, "Low-dose CT image denoising using a generative adversarial network with Wasserstein distance and perceptual loss." *IEEE Transactions on Medical Imaging*, 37, no. 6, 1348–1357, 2018.

16. Chauhan, Prachi, Hardwari Lal Mandoria, Alok Negi, and R. S. Rajput. "Plant diseases concept in smart agriculture using deep learning." *Smart agricultural services using deep learning, big data, and IoT*, pp. 139–153. IGI Global, Hershey, PA, 2020.

17. Ma, Xingjun, Yuhao Niu, Lin Gu, Yisen Wang, Yitian Zhao, James Bailey, and Feng Lu. "Understanding adversarial attacks on deep learning based medical image analysis systems." *Pattern Recognition* 110, 107332, 2020.

18. Bhardwaj, Akashdeep, Fadi Al-Turjman, Manoj Kumar, Thompson Stephan, and Leonardo Mostarda. "Capturing-the-invisible (CTI): Behavior-based attacks recognition in IoT-oriented industrial control systems." *IEEE Access* 8, 104956–104966, 2020.

19. Alshehri, Mohammed, Akashdeep Bharadwaj, Manoj Kumar, Shailendra Mishra, and Jayadev Gyani. "Cloud and IoT based smart architecture for desalination water treatment." *Environmental Research* 195, 110812, 2021.

20. Gao, Yansong, Bao Gia Doan, Zhi Zhang, Siqi Ma, Anmin Fu, Surya Nepal, and Hyoungshick Kim. "Backdoor attacks and countermeasures on deep learning: A comprehensive review." arXiv preprint arXiv:2007.10760, 2020.

21. Yuan, Xiaoyong, Pan He, Qile Zhu, and Xiaolin Li. "Adversarial examples: Attacks and defenses for deep learning." *IEEE Transactions on Neural Networks and Learning Systems* 30, no. 9, 2805–2824, 2019.

22. Zhang, L, Lin G, Gao B, Qin Z, Tai Y, Zhang J Zhang, Liqiang, Guanjun Lin, Bixuan Gao, Zhibao Qin, Yonghang Tai, and Jun Zhang. "Neural model stealing attack to smart mobile device on intelligent medical platform." *Wireless Communications and Mobile Computing* 2020, 10, 2020.

3 DNA-Based Cryptosystem for Connected Objects and IoT Security

Brahim Belhorma
High School of Signals (HSS), Koléa, Algeria

Mounir Bouhedda and Billel Bengherbia
University of Medea, Médéa, Algeria

Omar Benzineb
University of Blida, Blida, Algeria

CONTENTS

DOI: 10.1201/9781003140023-3

3.1   INTRODUCTION

The IoT is starting to gain popularity. Through linking all possible things to the internet, from cameras and phones to fridges to wind turbines, people are finding new ways to use IoT in order to improve the quality of life.

The WSN was initially designed as a closed network as a very common concept for the tracking of physical data and environmental conditions. The ambient information was gathered via sensor nodes and sent via a gateway to a remote place. In such a layout, there are no direct links between the end-users and the sensor nodes. Both signals are sent through the gateway between nodes and the outside world. For example, the use of sensors for the detection of enemy attacks on the field was inspired by military applications. The IoT is an extension of WSN, an expansion of the traditional model of the internet in which the digital world reaches the physical world.

Nevertheless, IoT is opening up a completely new aspect to security threats. Indeed, IoT provides compatibility for both forms of communication: human-to-machine and machine-to-machine. All are expected to be fitted with small, embedded devices that can connect to the internet in the near future. This skill is beneficial in many areas of our everyday lives: from building automation, smart cities, and monitoring systems to all portable smart devices. However, the more IoT applications are deployed, the greater the risk to our information system. Nonetheless, the undenied number of IoT systems, due to limited infrastructure and lack of defense mechanisms, is vulnerable to security breaches, for example, denial of service and replay attacks. IoT systems have many security challenges to overcome in order to obtain the necessary security.

A new encryption and decryption process based on DNA structures is introduced in this chapter, molecular characteristics and biological functions in addition to Huffman compression, which is used to reduce transmitted data size. The encryption method is conceptually based on the transcription and translation cycle. They refer to the two biological DNA operations: replication and transformation of DNA into a protein. The cryptographic DNA algorithm is developed and applied mainly at physical but not molecular levels. A cryptographic algorithm is developed and applied in this chapter for key generation and message encryption. The generated keys are validated using cryptography standard tests which demonstrate their very good performance.

To validate the proposed approach, Arduino MEGA Boards are used with NRF24L01 radiofrequency transceivers to build a WSN-based hardware platform. An HMI based on IoT using Node-Red of IBM, hosted in a Raspberry Pi 3, is created to monitor the collected data or to control actuators from a specific remote place securely.

3.2   RELATED WORKS

The development of cryptographic algorithms has seen considerable success in the last decades. The early inventions were monoalphabetic ciphers, multialphabet replacements, transposition ciphers, and block cipher (Paar et al. 2010). Additional encoding algorithms like AES, DES, RSA, and SHA followed (Preneel 2007; Rubinstein-Salzedo 2018).

Even so, future demands predict that vast collections of data and memory efficiency can also be encrypted using DNA methods (Tornea and Borda 2009). In 2003, Chen (2003) proposed the molecular theory-based cryptographic DNA, One-Time-Pad (OTP), and performed encryption/decryption of the 2D image. In 2013, Tornea and Borda (2013) proposed DNA-based index encryption. Biswas et al. (2019), propose a technique for DNA cryptography based on dynamic mechanisms. Chai et al. implemented complex DNA encryption and chaos cryptosystem for a colored picture (Chai et al. 2019). In 2012 Sangwan published a paper about a combination of a new Symmetric-key algorithm and Huffman compression algorithm for text encryption (Sangwan 2012). A chaotic encryption system to meet the security need on WSN using chaotic systems was carried out (Bayilmis et al. 2017). However, the proposed approach combines the DNA methods for text encryption using Huffman compression for size optimization. An experimental implementation and cryptoanalysis tests are done for tests and validation.

3.3   THEORY AND BACKGROUND

3.3.1   CRYPTOGRAPHY

Public contact institutions share information on secrecy, for example, financial transactions, medical reports, and personal records. An accidental receiver poses a significant danger to the confidentiality of classified information. Cryptographic strategies help to ensure that confidential information is safe. Cryptography helps the sender to store secret information safely or transfer it across vulnerable networks such that only the target destination is able to interpret it. An encrypted device encrypts data and creates an encrypted output that makes little sense to an unintentional consumer with no information regarding the key. For decryption, knowledge of the key is vital.

Cryptography may be commonly defined as symmetrical and asymmetrical. The two categories differ in the cryptography key where symmetrical encryption uses the same encryption and decryption key (Figure 3.1), contrary to the asymmetric contrary to the asymmetric cryptography, which uses different keys (Figure 3.2).

The reliability of encrypted data is primarily dependent on the strength of the encryption and decryption key. The legitimate user is able to decode the ciphertext in secure encryption systems using some available private information. However, it

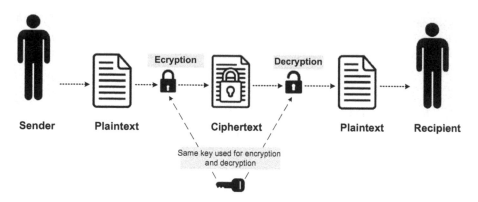

FIGURE 3.1   Symmetric cryptography.

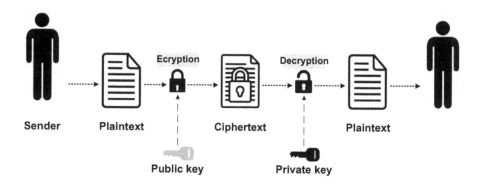

FIGURE 3.2   Asymmetric cryptography.

should be difficult for an attacker to decode the ciphertext that does not possess this private information. While cryptography allows sensitive information to be guarded, key breakers have introduced many cryptographic device cracking methods (Loukas 2015; Fink et al. 2017). DNA computing is a potential technology in the field of cryptography which brings new confidence for unbreakable algorithms, as traditional cryptography techniques based on mathematical and theoretical patterns may be at risk of intrusion (Zhang and Fu 2012; Cui et al. 2014; Suyel and Ganesh 2018).

3.3.2   DNA-Based Cryptography

DNA Computation is a new technology for protecting data using the biological structure of DNA. It was developed by Leonard Max Adleman in 1994 to solve difficult problems like the Hamilton path problem and the NP-complete problem like the Traveling Salesman Problem (TSP). The technique was later applied to encrypt and minimize the size of stored data, which allowed a quicker and more reliable data transfer over the network.

DNA strands are long polymers containing millions of nucleotides attached together. These nucleotides make one of four nitrogen bases, five-carbon sugar groups, and one phosphate group. The nucleotides that constitute these polymers originate from: cytosine (C), adenine (A), guanine (G), and thymine (T). This implies, mathematically, that four letters {A, G, C, T} can be used for encoding information, which is more than satisfactory since the computer requires only two digits (1 and 0) for the same purpose.

In terms of the DNA chain, DNA cryptography can be described as a method of hiding data. In the cryptographic process, every character is transformed into a different combination of the four nucleotide bases of DNA.

DNA cryptography is an important evolving technique that operates on DNA computing concepts. Inside the microscopic nuclei of living cells, a large amount of information is handled by DNA. It encodes all the necessary instructions for every living creature. The main advantages of DNA computing are the miniaturization and parallelism of traditional silicon-based machines. For example, about a million transistors can support a square centimeter of silicon, while current manipulation techniques can handle 1020 DNA strands.

The use of nucleotide sequences in DNA has been suggested by many researchers for this intention (A for 00, C for 01, G for 10, T for 11). One concept is not only that the details will not be encrypted, they may even be buried in the DNA, and so the technique called DNA steganography is well protected. DNA has a huge capacity range of data storage. To have just a simplified idea, we can say the following three specifications about DNA:

- Ultra-compact recording media: Very large volumes of files are stored in a compact volume.
- One gram of DNA is made up of 1021 DNA bases (a size of Terabytes).
- The whole data stored in the world can be contained in a few grams of DNA.

3.3.3 HUFFMAN COMPRESSION

In 1952, David Huffman introduced a mathematical method to compact a binary code word with different symbols. For all symbols, the length of a word code is not identical: the most frequently appearing symbols are coded with small code words, while the rare symbols have long binary codes. This is called the prefixed Variable Code Length (VLC) to denote this encoding type since no other code is another code prefix. The final stream length would therefore be smaller than with constant-size encoding (Salomon 2008).

The Huffman coder produces a tree with all symbols and their presence frequency. Branches are recurrently constructed from the less common symbols (Sangwan 2012).

Building the tree is achieved by first ordering the symbols according to the frequency of appearance. The symbols with the lowest appearance frequency are then excluded from the list and added to a node whose weight is the sum of the two symbols' frequencies. By considering each node formed as a new symbol, the lower weight symbol is assigned to branch 1, the other to branch 0, and so on, until there is a single parent node called the root. Each symbol's code corresponds to the code sequence along the path that runs from it to root. Therefore, the deeper symbol in the tree would have a longer code (Sayood 2012).

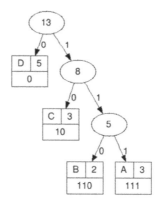

FIGURE 3.3   Huffman tree of the expression "DAADCCBCBDDAD."

TABLE 3.1
Results of Huffman Compression for the Expression "DAADCCBCBDDAD"

Character	Code	Frequency	Size
D	0	5	$5 \times 1 = 5$
C	10	3	$3 \times 2 = 6$
A	111	3	$3 \times 3 = 9$
B	110	2	$3 \times 2 = 6$
$4 \times 8 = 32$ bits		13 bits	26 bits

As an example, the Huffman tree of the expression "DAADCCBCBDDAD" is given by Figure 3.3, when we can notice rectangles divided into three parts containing the character, its frequency, and the corresponding Huffman code (Table 3.1). This method of encoding compression gives a good compression rate, particularly in text and monochrome images (faxes, for example).

Without encoding, the total string size was $13 \times 8 = 104$ bits. After Huffman encoding, the string size is reduced to $32 + 26 + 13 = 71$ bits. For decoding, the Huffman code is used to browse the tree to find the character.

3.4   PROPOSED CRYPTOSYSTEM-BASED DNA

In this section, the contribution is presented in terms of encryption/decryption, and the detail of the technical and practical implementation of hardware and programming solutions is given. This includes setting up a cryptosystem, in a small WSN, to send DNA-based encrypted messages from a sender node to a recipient node that can decipher the encrypted received message. To validate the performed work, an electronic implementation is performed for testing and analysis of the developed algorithm.

3.4.1 SPECIFICATIONS PRESENTATION

In order to realize the proposed cryptosystem, requirements are outlined, making the four main objectives of this work clear and precise which are namely:

- Presentation and explanation of the encryption and decryption algorithms-based DNA.
- Analysis and testing of the algorithm.
- Presentation of the corresponding electronic implementation.
- Presentation of the designed HMI using Node-Red in the context of IoT.

3.4.2 ENCRYPTION PROCESS

The main purpose of DNA cryptography is to encrypt the text in plain text and hide it in digital DNA form. This ensures higher data confidentiality where the key can be generated and modified randomly and adaptively in the function of data length.

3.4.2.1 Consideration for the Key Generation

The DNA-encrypted sent message uses a key based on the genes of the famous Algerian animal "Fennec Fox" (Vulpes zerda) as the encryption base (NCBI 2014). Table 3.2 gives the first 1080 genes. In the remainder of this section, we will give examples of the encryption/decryption of the word "**help**."

3.4.2.2 Phases of Encryption Process

In the proposed encryption algorithm, nine phases are followed to generate the encrypted message. It can be summarized by the flowchart of Figure 3.4, where each phase is explained separately in detail in the following sections.

- *Phase 1: Huffman coding of clear text*
  It corresponds to the compression of the clear text to be sent using Huffman's encoding, where only 67 characters are used in order to optimize the text size. This encoding allows, in addition to the compression of the clear message, to make a first encryption of the cleartext.

    Result → Huffman code of the text to be encrypted = 20$_c@X

- *Phase 2: Generation of Plain Text Codons*
  Each character obtained in the Huffman code from the clear text is converted to its corresponding DNA code according to Table 3.3.

    Result → Codons = CAATGATCGGAATTAGGATGATGTACGT

- *Phase 3: Generation of the encryption key*
  The DNA gene used to generate the key is a part of 1080 chromosomes long from the animal Fox Fennec (Vulpes zerda). Two parameters (P1 and P2) are randomly generated to create this key. P1 is the position of the first chromosome, and P2 is the length of the key. Of course, the generated key takes into account the length of the message to be encrypted.

TABLE 3.2
First 1080 Genes of "Fennec Fox" (Vulpes zerda) (NCBI 2014)

Gene	Fennec Fox (Vulpes zerda) Genes					
1–60	GTTAATGTAG	CTTAATTAGT	AAAGCAAGGC	ACTGAAAATG	CCAAGATGAG	TCATAAGACT
61–120	CCATAAACAC	AAAGGTTTGG	TCCTGGCCTT	CCTATTAGTC	CTTAGTAAAC	TTACACATGC
121–180	AAGCCTCCAC	GCCCCAGTGA	GAATGCCCTT	AAAATCATTA	ATGATCTAAA	GGAGCAGGTA
181–240	TCAGGCACAC	TCCCTAGTAG	CCCATAACAC	CTTGCTAAGC	CACGCCCCCA	CGGGACACAG
241–300	CAGTGATAAA	AATTAAGCCA	TGAACGAAAG	TTCGACTTAG	TTACACTAAA	GAGGGTTGGT
301–360	AAATTTCGTG	CCAGCCACCG	CGGTCATACG	ATTAACCCAA	ACTAATAGGC	CAACGGCGTA
361–420	AAGCGTGTTT	AAGATGACAC	AACACTAAAG	TTAAAACTTA	ACTAAGCCGT	AAAAAGCTAC
421–480	AGTTACAATA	AAATATGCTA	CGAAAGTGAC	TTTAAAATTT	TCTGACTACA	CGATAGCTAA
481–540	GACCCAAACT	GGGATTAGAT	ACCCCACTAT	GCTTAGCCCT	AAACATAAAT	AGTTGCGCAA
541–600	CAAAACAATT	CGCCAGAGAA	CTACTAGCAA	CAGCTTAAAA	CTCAAAGGAC	TTGGCGGTGC
601–660	TTTATATCCC	TCTAGAGGAG	CCTGTTCTAT	AATCGATAAA	CCCCGATAAA	CCTCACCATC
661–720	CCTTGCTAAT	ACAGTCTATA	TACCGCCATC	TTCAGCAAAC	CCTTAAAAGG	CAGAATAGTA
721–780	AGCAAAATCA	TCACGCATAA	AAAAGTTAGG	TCAAGGTGTA	ACTTATGGGA	TGGGAAGAAA
781–840	TGGGCTACAT	TTTCTATTTT	AAGAATATTC	TACGAAAGTT	TTTATGAAAC	TAAAAACTGA
841–900	AGGAGGATTT	AGTAGTAAAT	TAAGAATAGA	GAGCTTAATT	GAATAGGGCC	ATGAAGCACG
901–960	CACACACCGC	CCGTCACCCT	CCTCAAGTAA	TAAGACTGAG	ACCACAATCA	TATTAACTCA
961–1020	TACCAAAACA	CGAGAGGAGA	TAAGTCGTAA	CAAGGTAAGC	ATACCGGAAG	GTGTGCTTGG
1021–1080	ATCAACCAAA	GTGTAGCTTA	ATAAAAGCAT	CTGGCTTACA	CCCAGAAGAT	TTCATAACTA

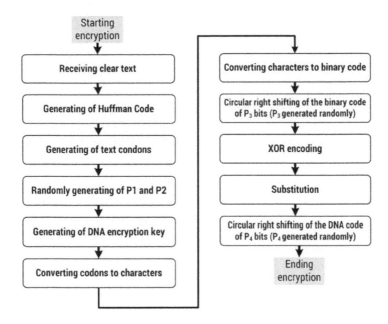

FIGURE 3.4 Flowchart of the encryption process.

TABLE 3.3
Characters' Codons

Codons	Character	Codons	Character	Codons	Character	Codons	Character
CGGG	(	GAAC	9	GACC	I	GAGC	Z
CGCA	)	TCAC	:	CATG	G	TCGG	[
TCAA	*	ACTC	;	CGTA	K	GACG	]
TTCC	+	GGCA	<	TTGT	L	TCGC	^
ATTG	,	GGGG	=	ACGA	M	TTAG	_
GAGG	-	TCGA	>	TTCT	N	GACA	'
ACTT	.	ATCC	?	TTTG	O	GAGT	a
TCCT	/	ATGT	@	CGGC	P	TTAC	b
GATC	0	ACCC	A	ATGC	Q	GATG	c
CGCG	1	GAAT	B	GGAT	R	CATC	d
CAAT	2	GGTC	C	CACG	S	CATT	e
CAAC	3	ACGG	D	CGAC	T	TCTC	f
TCCG	4	CGCT	E	CAGC	U	TTCG	g
TCAG	5	ACAT	F	ACAC	V	ATAG	h
GAGA	6	CGAG	G	ACGT	W	GGTT	i
CGAA	7	ATTA	H	TCCA	X	GGAA	$
CATA	8	GGCG	I	CAAA	Y	-	-

TABLE 3.4

Encoding Two DNA Bases in English Capital Letters

The DNA chain agreed in the key	T	G	A	C	T	G	A	C	T	G	A	C	T	G	A	C
The DNA chain resulting from the Huffman code	A	A	A	A	C	C	C	C	T	T	T	T	G	G	G	G
Encoded character	D	I	R	P	X	W	Z	F	B	H	K	M	S	V	Y	E

Result → P1 = 274, P2 = 22, Key = AATGAACTGAAAAAGGCCTAGA

- *Phase 4: Conversion of Codons to Characters*
  Now, two chains of DNA are obtained. The first chain corresponds to the compressed plain text coded in DNA base (codons) obtained in phase 2, and the second chain corresponds to the key generated in phase 3. By combining these two chains and according to Table 3.4, a message which contains characters in uppercase is gotten.

Result → Codons in characters =
ZRDHYRMXVYRRKKIVEPBYIKYKDWYK

- *Phase 5: Binary Coding*
  The string obtained in phase 4 is converted into a binary code where each character is replaced by its code, according to Table 3.5. The result is a binary data frame.

Result → Binary code = 11010100100010001100110010011011100010110
11011001100101001001011010110100110110001010110010100101011110
0101011001001011111100111

- *Phase 6: Circular shift right process*
  This is a circular shift right of P3 bits of the code obtained in phase 5. P3 is a value generated randomly.

Result → P3 = 4, Binary code = 10111101010010001000110011001001101
11000101101101100110010100100101101011010011011000101100000001
0110010100101011110010101100100101111110010

- *Phase 7: XOR Coding*
  The message obtained in phase 6 with the DNA-generated key is converted into binary code according to Table 3.6 bit to bit with an XOR gate.

Result → Binary code after XOR gate with encryption key = 10110011010
01111000010001100011011000111000010111000011000101010010110100
111011010101010110011100101111011001010101111100011101110110 1111
11100

- *Phase 8: Substitution*
  In this step, a substitution of each bit pair of the code obtained in phase 7 is performed into a nucleotide base according to the encoding given by Table 3.6.

TABLE 3.5
Adopted Capital Letters Codes

Letter	B	D	E	F	H	I	K	M	P	R	S	V	W	X	Y	Z
Sequence	2	4	5	6	8	9	11	13	16	18	19	22	23	24	25	26
Code	00010	00100	00101	00110	01000	01001	01011	01101	10000	10010	10011	10110	10111	11000	11001	11010

TABLE 3.6
Binary Codes of Nucleotide Bases

DNA base	A	C	G	T
Binary code	00	01	10	11

Since each bit pair is converted into a single character that represents a nucleotide base, a DNA code is obtained at the end by adding the genetic code of the randomly generated parameters (P1, P2, and P3).

Result → DNA Code = GTATCATTAAGAGTACTAAGTGACG AGGGCCGGCTCGGGGGTATGCCTGTAGGTTGAT GTCTTTACGTCCGGATCCAACCGCGGATCTCCG

- *Phase 9: Circular right shift process*
  To enhance the security of the algorithm, a circular right shift of P4 bits is done to the DNA code generated in phase 8. P4 is an integer value generated randomly whose genetic code is added at the end of the DNA code.

Result → P4 = 4, Final Encrypted Text = TCCGGTATCATTAAGATAC GTACTAAGTGACGAGGGCCGGCTCGGGGGTATGCCT GTAGGTTGATGTGTCTTTACGCGTCCGGATCCA ACCGCGGATCTCCG

To show the advantage of the proposed method by running the algorithm several times, we note the achievement of totally different results for the encryption of the same clear text (Table 3.7).

TABLE 3.7
Five Times of Algorithm Execution Result for the Encryption of the Same Clear Text

Execution Number	Ciphertext
01	ACATACAACTGTCATGTTTGCCTTAGTGGCGTTACGGCTGGAGTAAGGTCT TGGCTAGTCGAGGATTACGTCTAGCTAGGATCCAACGAGACAACGAGGAAC
02	AGCATAAGACGCCGTAAGCACGTGTACTTGCGGTATCCGTAAACATACAAGA GAAAGGTATCAAGTTACATACATGCGCGTCAGCAACGAGAGATCTCGAGA
03	TCCGTCGTTATCCTTTCGGTTCGGTTTTATGTCATTCGTCCCCGTGCGGTTCG GCACCTTTCGTAGCCGTTTTCCAATTCAGCGAAGATCCAATCGAATCCG
04	GAGATCCGGGTGTACTTTAACAGTCGCGCCCCGTACCACTCGTGGTCGCATT CACTATGATTCAATCTTTTTACTTGTGATCCGAACGCGCGCGCGAACATA
05	TACAACTCATCCTTATCTGGATGCGGGCTCCTTTGACAACTAACTTCCAGT GGCAGGGAATTAGTATGAAACTGGGGATCCAACGATCCAATCAATCAGAGA

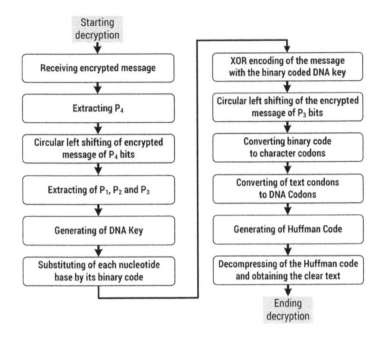

FIGURE 3.5 Flowchart of the decryption process.

3.4.3 Decryption Process

The decryption process is carried out in a reverse way to the encryption process. Note that the four random values generated during the encryption process are sent securely with the ciphertext in order to generate the encryption DNA key and make the necessary circular rotations. The flow chart given in Figure 3.5 summarizes the necessary steps for the decryption process.

3.4.4 Security Evaluation

3.4.4.1 Frequency Analysis

The encrypted text histogram, if securely encrypted, has a uniform random distribution, which means that the encrypted text values were randomly generated from a uniform distribution (Paar et al. 2010). The first problem with "classic" encryption is its vulnerability to frequency analysis. The English alphabet is used in classical encryption, so it is obvious that the correct letter of the encrypted text can be guessed using frequency analysis. For example, the letter that occurs most frequently in the encrypted text should be the letter "e," which is the most frequent letter in the English language. To decrypt the encrypted text, this strategy can be adopted. The vulnerability against frequency analysis is overcome in the proposed DNA-based cryptosystem. In this algorithm, Huffman coding compresses the text and then transforms it

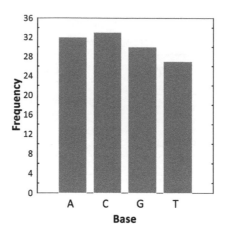

FIGURE 3.6   Bases frequencies of the ciphertext.

into DNA chains, converted to binary and then to DNA chains. So, the frequency of letters should not be guessed.

It can be seen that the letter "e" is the most common in the text to be encrypted. However, any relationship can be found with the other letters to guess the clear text from the ciphertext. Figure 3.6 gives the frequency of each nucleotide base of the encrypted message relating to the clear message "send help soon." There is almost a similarity in the frequency of each base.

3.4.4.2   Encryption Key Security Analysis

The replication of its secret key, where key length is limited, is another popular problem with classical encryption. When the key length is determined, the entire encryption text is broken since before, and after each section, the length of each section is specified to extract clear text from the frequency analysis approach. The main test for finding the right key length is the Friedman test (Wang, Tamaki, and Curty 2018). Friedman's test uses a coincidence index that measures the similarity of the encryption letters to crack the encryption key. The Friedman test is intended to evaluate the key's length (KL) that can be expressed by relation (3.1).

$$KL = \frac{K_p - K_r}{K_0 - K_r} \tag{3.1}$$

With

K_p: the probability that two numerical elements are identical. In English, this value is 0.067 (Ferguson, Schneier, and Kohno 2010).

K_r: coincidence probability of a random alphabet collection. Four letters A, C, T, and G are used for DNA coding. This value, therefore, amounts to 0.25.

K_0: is the coincidence rate. It is given by the relationship (3.2) (Ferguson, Schneier and Kohno 2010).

$$K_0 = \frac{\sum_{i=1}^{c}\left(f_i \times f_{i-1}\right)}{N(N-1)} \tag{3.2}$$

With c: the size of the alphabet, N: the length of the encrypted text, f_i is the letter frequency. We calculate this value is calculated for the encryption of the clear text "send help soon." We have the data given by expression (3.3).

$$\sum_{i=1}^{c}\left(f_i \times f_{i-1}\right) = A:\left(32\times31\right)+C:\left(33\times32\right)+G:\left(30\times29\right)+T:\left(27\times26\right) = 3620 \tag{3.3}$$

$N = 122$, $N(N-1) = 122 \times 121 = 14762$ from which

$$KL = \frac{K_p - K_r}{K_0 - K_r} = \frac{0.067 - 0.25}{0.2452 - 0.25}, \text{ consequently, } KL = \frac{K_p - K_r}{K_0 - K_r} = \frac{0.067 - 0.25}{0.2452 - 0.25}$$

It can be noticed that the denominator of KL is almost zero, which can give a long length of a key for a long ciphertext. This makes it very difficult to find clear text and encryption keys (Rubinstein-Salzedo 2018).

3.4.4.3 Entropy of the Encryption Key

Entropy measures the uncertainty or randomness of the key. With the chosen key that corresponds to a portion of the Fennec Fox gene of 512 values and with the conditions adopted in the encryption algorithm, 129,792 keys can be generated. The result is that the probability of obtaining the key is $1/129,792 = 7.7 \times 10^{-6} \cong 0$. Entropy is almost equal to zero, which means that there is no certainty about the prediction of the key (Preneel 2007).

3.5 CRYPTOSYSTEM HARDWARE IMPLEMENTATION

For the validation of the encryption and decryption process, the encryption method is implemented on a system based on microcontrollers and radio frequency transmission modules.

3.5.1 GENERAL DESCRIPTION OF THE CRYPTOSYSTEM

The system uses a two-node wireless network (Encryption and decryption nodes!). A node captures the data and sends it encrypted to the second node for decryption. Figure 3.7 shows the synoptic diagram of the adopted electronic assembly for the encryption/decryption. An Arduino MEGA 2560 microcontroller is used for each node. The NRF24L01 modules are connected to each node for wireless transmission and reception. The Arduino MEGA microcontroller for node 1 contains the encryption and emission algorithm written in C language. Similarly, the decryption and reception program is implemented in the Arduino MEGA microcontroller for node 2.

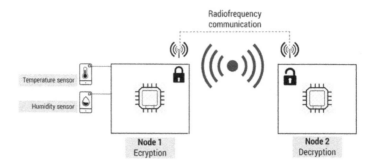

FIGURE 3.7   Synoptic diagram of the adopted system.

3.5.2   Presentation of Used Components

3.5.2.1   Temperature and Humidity Sensor DHT11

The DHT11 sensor is a temperature and humidity sensor "two in one." The DHT11 is capable of measuring temperatures from 0 to + 50°C with +/− 2°C accuracy and 20%–80% relative humidity with +/− 5% accuracy. A measurement can be performed every one second (Jeremy 2013). The DHT11 (Figure 3.8) is compatible with 3.3 volts and 5 volts (however, the manufacturer still recommends that the sensor be supplied with 5 volts for accurate measurements). DHT11 cannot measure (and withstand) negative temperatures or temperatures above 50°C. The DHT11 sensor consists of three pins: GND, + Vcc, and data. Figure 3.9 shows the connecting scheme of DHT11 to Arduino MEGA.

FIGURE 3.8   DHT11 sensor.

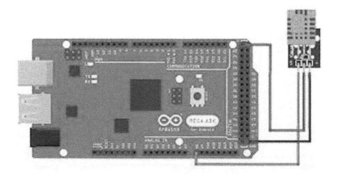

FIGURE 3.9   Connecting DHT11 to Arduino MEGA.

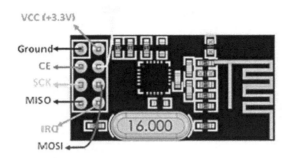

FIGURE 3.10   NRF24L01 module.

3.5.2.2   Communication Radio Module NRF24L01

The NRF24L01 radio module (Figure 3.10) is a fully integrated radio module from the Nordic Semi-Conductor manufacturer. It is a radio module that integrates everything needed to transmit and receive data over the 2.4 GHz frequency range (such as Wi-Fi or Bluetooth) using Nordic's proprietary communication protocol called "ShockBurst." This communication protocol allows the NRF24L01 to be considered as a complete modem, with addressing, handling of transmission errors, and automatic retransmission in case of non-response of the recipient.

The maximum flow of the nRF24L01 + is 2Mbps, but at this speed, the range is only a few meters. The nRF24L01 + can transmit or receive data of 32 bytes per communication channel, for a total of six simultaneous communication channels. The maximum size of a packet is 32 bytes. The nRF24L01 radio module communicates with its master via an SPI bus. The three spindles common to the SPI bus (Figure IV-7) can therefore be found at the pinout level.

- MISO: Send slave data to the master.
- MOSI: Sends data from the master to the slave.
- SCK: clock.
- CSN: switch between the transmit mode and the receive mode.
- CE: switch from active mode to standby mode.
- VDC: supply voltage between 1.9 and 3.6v.
- GND: connection to ground.

3.5.2.3   Mounting Principle (Transmitter/Receiver)

The pin-in of the transmitter and receiver with Arduino MEGA 2560 is shown in Figure 3.11. The NRF24L01 has a maximum sending capacity of 32 bytes. Indeed, for an encrypted message exceeding this capacity, it must be divided into 32-byte packets and finish the last package by an "F" character to indicate to the receiver the end of the message where he must gather it for decryption. Figures 3.12 and 3.13 show the encryption and decryption results from the two microcontrollers corresponding to node 1 and node 2, respectively.

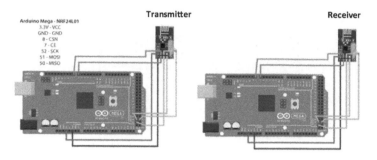

FIGURE 3.11   Transmitter/receiver pin-in with Arduino MEGA 2560 and NRF24L01.

FIGURE 3.12   Encryption result of node 1 (COM6).

FIGURE 3.13   Decryption result of node 2 (COM4).

3.6   HUMAN-MACHINE INTERFACE (HMI)

In this project, the HMI consists of a web application accessible from a browser. It is not always easy to program the link between the hardware (sensors) and the web page provided to the user. Combined with a hardware solution consisting of a Raspberry and Arduino boards, The Node-Red platform of IBM proves to be a very interesting solution since it is an "open source" application, and in addition, it has a large user community.

Node-RED is a powerful tool for developing IoT applications to simplify programming using "nodes" to perform tasks through predefined code blocks. It is a visual programming tool that enables developers to link code blocks together. Linked nodes, typically a set of input nodes, processing nodes, and output nodes, constitute

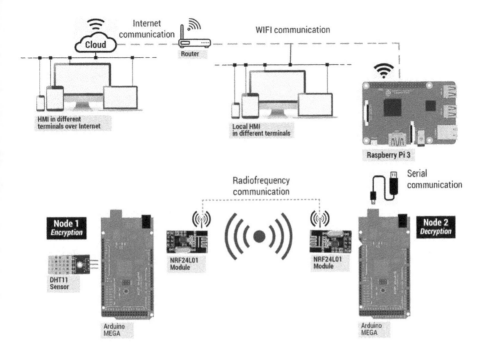

FIGURE 3.14 Overall architecture of the proposed system.

a "flow" when wired together. The programming interface is accessed through a browser. It is better to use the browser of a computer connected to the same network as the Raspberry. The URL that allows us to access the programming interface is http://ip_raspberry:1880.

3.6.1 TRANSFER OF DATA ACQUIRED BY SENSORS

Once the sensor's data are available at node 1, they must be transferred to the Raspberry to be displayed locally and in the cloud. The data transfer is done via the dev/ttyACM0 serial port. The proposed monitoring and control system is illustrated in Figure 3.14, where the different nodes around Raspberry and the connection between them are illustrated.

3.6.2 VISUAL PROGRAMMING OF THE HMI

3.6.2.1 Splitting Data

Data from node 2 board reaches the Raspberry via the serial port usually named/ dev/ttyACM0. Therefore, a serial node is used on the Node-Red flow and configured for a connection to this port. The serial communication offers a data frame composed of temperature and humidity. Splitting these data is necessary for displaying them separately, for this purpose, a predefined processing function "Json" is used (Figure 3.15).

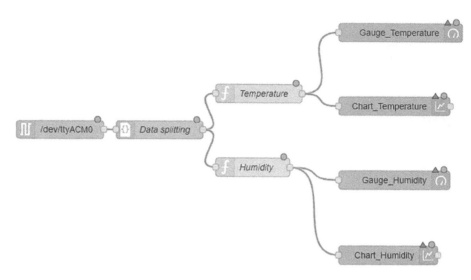

FIGURE 3.15   Temperature and humidity display Node-Red flow.

3.6.2.2   Temperature Display

Two components are used to show the temperature and humidity variation. For these components to be properly arranged on the web page, they must be placed in a Tab (Visualization), then in a Group (Temperature or Humidity), in the group selection box, a graph and a gauge are configured to display the temperature and humidity (Figure 3.15).

3.6.3   HMI VISUALIZATION

As the visual programming of the HMI is finished, the program is validated by the deploy button, then the HMI can be displayed on a website accessed by the URL: http://ip_raspberry:1880/ui. The interface achieved in our project is illustrated in Figure 3.16, where the values of temperature and humidity can be clearly displayed in graphic and gauge mode.

3.7   IoT-BASED SUPERVISION

There are cloud instances for the supervision and control of connected objects, for this purpose, FRED (Front End for Node-Red) can be used.

3.7.1   FRED (FRONT END FOR NODE-RED)

FRED is a tool of Node-Red adopted by several online users, FRED offers a free account, but with some limitations for its free version. FRED is used to share the information collected in the cloud and control the different actuators from the internet. The following steps must be followed in order to use FRED:

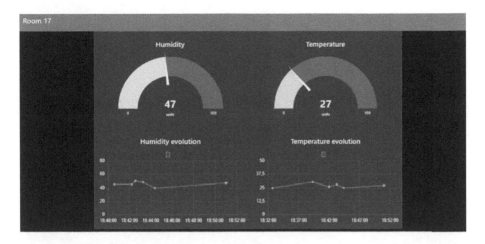

FIGURE 3.16 Temperature and humidity display in the HMI.

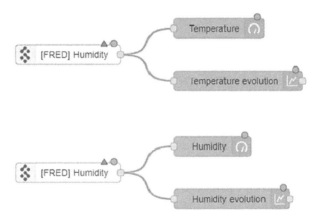

FIGURE 3.17 Cloud data sharing using FRED.

- Installing the input/output function (FRED) on Node-Red.
- Creating an STS (Sense Tecnic Services) account on https://fred.sensetecnic.com/
- Configuring data exchange on Node-Red/cloud.

Sensor data are sent to the cloud using the (FRED output) output function on the visual programming environment (Figure 3.17). The input function (Fred input) is used and connected to the opened STS account to display sensor data (Figure 3.18).

3.7.2 VISUALIZATION OF HMI ON THE CLOUD

The input/output FRED functions must be connected to the flow. The website can be accessed using the URL: https://work26.fred.sensetecnic.com/api/ui/, where "work26" is the name given to the project on FRED. Figure 3.19 shows the interface

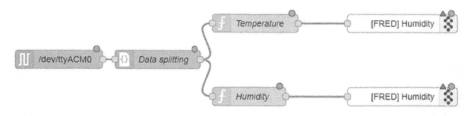

FIGURE 3.18   Displaying data on the cloud using FRED.

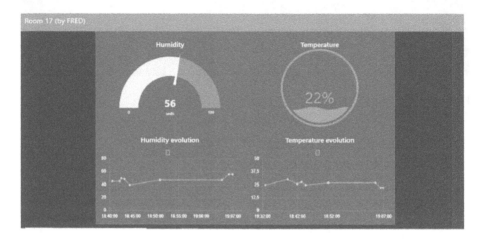

FIGURE 3.19   HMI on the cloud.

obtained from the cloud with different IP addresses through a laptop and a smart-phone. Of course, the HMI can be accessed via a dedicated website for our applica-tion. This requires configuration at the modem level where the system is connected as well as the Raspberry. This configuration will allow access to the HMI via an external IP address when the security of the network must be taken into account.

3.8   CONCLUSION AND FUTURE WORK

The purpose of this work was to explore and create an intelligent security encryption scheme for connected objects and IoT "DNA-based cryptography." applied to two transmission nodes. To accomplish this task, a DNA-based encryption algorithm with Huffman coding is proposed. The developed algorithms are implemented around Arduino MEGA 2560 microcontrollers, radio transmitters/receivers, and a Raspberry Pi 3 board.

Based on these findings, a cryptographic approach based on DNA sequences is presented. The approach, in its initial phase, generates the Huffman code of clear text. In the second phase, a DNA key is extracted from the genes of the Fennec Fox, where the position and the length are generated randomly with each encryption. Many operations on the DNA code are made: encoding, substitution, and rotation of a random number of bits generated for each encryption. The approach was implemented and then validated. We highlight the value of our approach through experiments. The results obtained showed, on the one hand, the impact of the use of DNA sequences and, on the other hand, the robustness of the proposed key generation method. The proposed solution has several advantages:

- Secure communication where it is noted that three levels of security have been adopted, namely:
  - Huffman Coding.
  - Random generation of encryption key parameters at each encryption operation.
  - Security regarding the use of DNA coding operations.
  - Random generation of binary code rotation bit numbers twice. These values are generated with each encryption operation.
- Passing two levels of analysis successfully:
  - Frequency analysis: Obtaining practically the same frequency for all DNA codes,
  - Encryption Key Security Analysis: The length of the key tends to a great value according to Friedman's formula. Therefore, it is very difficult to find the key.
- Using wireless transmission using the NRF24L01 modules.
- Supervision locally and via the cloud.
- Supervision and control from multiple terminals (PC, smartphone, tablet).
- Implementation of an HMI for supervision.

As a perspective to our work, we can enrich our approach with new features such as the application of multimedia content like images and video. Additional protection and verification mechanisms to improve the proposed scheme will be used for future work in this direction. Also, in the world of cryptography, artificial intelligence has a wide application which can be a very good way to follow.

ACKNOWLEDGMENT

This research work is supported by the DGRSDT/MESRS (Directorate General of Scientific Research and Technological Development/Ministry of Higher Education and Scientific Research) of Algeria.

REFERENCES

Bayilmis, Cuneyt, Unal Cavusoglu, Akif Akgul, Sezgin Kacar, and Abdullah Sevin. 2017. "Enhanced Secure Data Transfer for WSN Using Chaotic-Based Encryption/Poboljsani Sigurni Prijenos Podataka Za WSN Uporabom Koda Zasnovanog Na Kaosu." *Tehnicki Vjesnik - Technical Gazette* 24 (4): 1065–1070.

Biswas, Md Rafiul, Kazi Md Rokibul Alam, Shinsuke Tamura, and Yasuhiko Morimoto. 2019. "A Technique for DNA Cryptography Based on Dynamic Mechanisms." *Journal of Information Security and Applications* 48 (October): 102363.

Chai, Xiuli, Xianglong Fu, Zhihua Gan, Yang Lu, and Yiran Chen. 2019. "A Color Image Cryptosystem Based on Dynamic DNA Encryption and Chaos." *Signal Processing* 155: 44–62.

Chen, Jie. 2003. *"A DNA-Based, Biomolecular Cryptography Design."* In *Proceedings – IEEE International Symposium on Circuits and Systems*. Vol. 3.

Cui, Guangzhao, Dong Han, Yan Wang, Yanfeng Wang, and Zicheng Wang. 2014. "An Improved Method of DNA Information Encryption." *Communications in Computer and Information Science*, 472, 73–77. Springer, Berlin.

Ferguson, Niels, Bruce Schneier, and Tadayoshi Kohno. 2010. *Cryptography Engineering: Design Principles and Practical Applications*. Wiley, Hoboken, NJ.

Fink, Glenn, Thomas Edgar, Theora Rice, Donald MacDonald, and Claire Crawford. 2017. "Security and Privacy in Cyber-Physical Systems." *Cyber-Physical Systems: Foundations, Principles and Applications*, 129–141. Elsevier, Amsterdam.

Jeremy, Blum. 2013. *Exploring Arduino: Tools and Techniques for Engineering Wizardry.* Wiley, Hoboken, NJ.

Loukas, George. 2015. "Physical-Cyber Attacks." *Cyber-Physical Attacks*, 221–253. Elsevier, Amsterdam.

NCBI. 2014. "Vulpes Zerda (ID 238103) – BioProject – NCBI." *Vulpes Zerda.* https://www.ncbi.nlm.nih.gov/bioproject/PRJNA238103.

Paar, Christof, Jan Pelzl, Christof Paar, and Jan Pelzl. 2010. "Introduction to Cryptography and Data Security." *Understanding Cryptography*, 1–27. Springer, Berlin.

Preneel, Bart. 2007. "An Introduction to Modern Cryptology." *The History of Information Security*, 565–592. Elsevier Science B.V., Amsterdam.

Rubinstein-Salzedo, Simon. 2018. *Cryptography.* Springer Undergraduate Mathematics Series. Springer International Publishing, Cham.

Salomon, David. 2008. *A Concise Introduction to Data Compression.* Undergraduate Topics in Computer Science. Springer, London.

Sangwan, Nigam. 2012. "Text Encryption with Huffman Compression." *International Journal of Computer Applications* 54 (6): 29–32.

Sayood, Khalid. 2012. *Introduction to Data Compression.* Elsevier, Amsterdam.

Suyel, Namasudra, and Chandra Deka Ganesh. 2018. *Advances of DNA Computing in Cryptography.* Chapman and Hall/CRC, Boca Raton, FL.

Tornea, Olga, and Monica Borda. 2009. *"DNA Cryptographic Algorithms."* In *International Conference on Advancements of Medicine and Health Care through Technology. IFMBE Proceedings*, Vol. 26, 223–226. Springer, Heidelberg.

Tornea, Olga, and Monica E. Borda. 2013. *"Security and Complexity of a DNA-Based Cipher."* In *Proceedings – RoEduNet IEEE International Conference.* IEEE, New York.

Wang, Weilong, Kiyoshi Tamaki, and Marcos Curty. 2018. "Finite-Key Security Analysis for Quantum Key Distribution with Leaky Sources." *New Journal of Physics* 20 (8): 083027.

Zhang, Yunpeng, and Liu He Bochen Fu. 2012. "Research on DNA Cryptography." *Applied Cryptography and Network Security.* InTech, Rijeka.

4 A Role of Digital Evidence
Mobile Forensics Data

G. Maria Jones
Saveetha Engineering College, Chennai, India

S. Godfrey Winster
SRM Institute of Science and Technology, Chennai, India

L. Ancy Geoferla
RMK Engineering College, Chennai, India

CONTENTS

DOI: 10.1201/9781003140023-4

4.1   INTRODUCTION

The computer, laptop, tablet, mobile devices, network connections have dramatically reshaped society. In the early stage of technology growth, i.e., 2 to 3 decades ago, it was difficult to own personal computers and mobile devices due to more expensive and challenging communication to others even through the mail. The internet connectivity was possible only through wire cables, modems, and users' needs to pay for access charges. Today, almost everybody in this world depends on mobile phones, computers, and the internet to carry out their activities. Many individuals own more than one laptop, and mobile devices are interconnected with the internet (Jones and Winster 2017). In addition to this, people have multiple email accounts for various usages like social media, personal, official, and business. Mobile devices have become the preferred devices for communication through texting, networking, and calling, which connect people soon. Indeed, the youngest generation tends to use social media networks, whereas the under 20 age group individuals are used to text messages rather than calling (Zickuhr 2010). The continuing innovation of technology creates opportunities for a good deed and an evil deed. There has been increasing technology growth in recent times where criminals used to develop new online criminal activities that did not even exist before. The internet and WWW (World Wide Web) provide a platform to perform crime and share information, and communicate with others.

4.1.1   Technology as Digital Evidence

Digital evidence is generally found in advanced devices, and it very well may be gotten from web-based media. Cell phones, PCs, laptops, and tablets are utilized to commit criminal activities saved and transferred to other electronic gadgets. The digital evidence can be found on a hard disk, cell phone, SD card, and some more. These are related to e-crimes like Mastercard misrepresentation, erotic entertainment, and different violations. Aside from this, occasionally, email likewise assumes a significant part by and large, which may have a lot of crucial proof identified with the case. For example, Bind, Torture, Kill (BTK) a serial killer who killed ten people from 1974 to 1991 and with the help of computer forensics tools, the floppy disk was analyzed and retrieved all the evidence which turned out to be a man named Dennis Rader (Landwehr & Landwehr n.d.). If child pornography, illegal drug communication is present in the mobile device or laptop, or computer, it is directed to the legal offense. This information is stored in digital form, and it can be either transferred to another digital medium. In every criminal activity on digital platforms, criminals may leave digital footprints after committing a crime. Even when criminals attempt to destroy, delete, modify the digital evidence, the forensics examiners can recover all detailed information with various methods. The forensics investigation tools are critical for analyzing complex cases lacking due to technical support and lack of advanced forensics software features in the last few decades.

4.1.2   Digital Forensic

Digital Forensics is the part of forensics science encompassing reconstructing, retrieving, extracting, identifying, preserving, and investigating the digital evidence

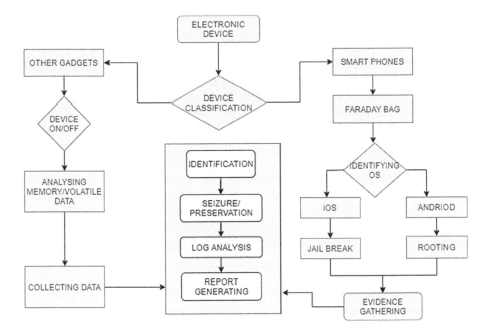

FIGURE 4.1 Work process to acquires data from the electronic device.

that resides in electronic devices. The main goal of forensics investigators (Figure 4.1) is to retrieve the old information from the device without modifying and altering it. Over the years, the digital forensics domain has rapid growth with various forms, which evolved as various sub-divisions like mobile forensics, IoT forensics, Cloud Forensics, Memory Forensics, Database Forensics, and many more. Ericsson's report stated that by 2022 the data traffic would reach 71 exabytes (Tamma et al. n.d.).

4.2 RELATED WORKS

The various existing works on analyzing digital forensics in a mobile environment are presented in this segment. Some of the techniques used in the existing system are discussed in this session: Songyang et al. (Wu et al. 2017) explored the WeChat application with some queries that came to the forefront during mobile forensics examination, and technical methods are proposed for answering the questions. Cosimo et al. (Anglano 2014; Anglano, Canonico, and Smartphones 2016) has analyzed WhatsApp and ChatSecure for reconstructing the chats and contacts from smartphones (statista) and also decrypted the database of all messages, and they concluded that none could reconstruct the data from the database by using the SQL cipher deletion technique. Anderson et al. (Rocha et al. 2017) proposed a method of authorship attribution applied to specific problems of social media forensics and also examined supervised learning methods for small samples with a step-by-step explanation.

Erhan et al. (Akbal et al. 2019) analyzed the BiP Messenger (BM) using forensics methodology for both rooted and unrooted devices, which showed the differences for data extraction. Dinil et al. (Divakaran et al. 2017) focused on the problem of evidence gathering by detecting malicious activities. Johannes et al. (Denzel, Stüttgen, and Vömel 2015) illustrated firmware manipulation techniques and proposed a method for identifying firmware-level threats in memory forensic investigations, and evaluated memory forensic tools within both physical and virtual environments. Shahzad et al. (Saleem, Popov, and Baggili 2016) (Jones and Winster 2017) evaluated two tools and showed that XRY dominates UFED in many cases in analyzing both performance and relevance. Vikram et al. (Harichandran et al. 2015) conducted a survey in cyber forensic to optimize resource allocation and to prioritize problems and possible solutions with more efficiency and finally obtained the results from 99 respondents who gave testimony which will be vital in the future. Robert et al. (Koch et al. 2015) proposed the degree and characteristics of logging based on geolocation. They gathered additional information and stored it in geo reputation for conspicuous locations. Since IP addresses are not static, additional information is also stored, and this information was used to reconstruct the attack routes, to identify and analyze distributed attacks.

4.3   MOBILE DEVICE FORENSICS

Mobile phones have become more sophisticated and predominantly in use by everyone. They started to evolve as smart devices with more advanced features and processing power, enabling them to connect with people across the globe. Since the smartphone provides more ease of access, but at the same time, it allows criminals to perform more criminal activities in the cyber world. Sometimes the illegal access of user's sensitive information like messages, contact lists, instant payment applications, confidential documents, emails, and so on can be stolen from mobile devices with advanced techniques. During an investigation, the information associated with mobile devices can help to provide an answer to find out the guilt. Mobile devices carry out some criminal activities like cyberstalking, cyber harassment, pornography, and many more activities for exchanging photos and videos. The majority of terrorist groups, drug dealers, are termed as an organized crime where mobile devices are used to share information and coordinate criminal activities (Casey and Turnbull 2011). So, data on mobile devices can provide many vital sources of evidence during an investigation.

Data acquiring from mobile devices are challenging due to its dynamic nature. Additional to this, the new launch of mobile devices and new models are devolved across the world. This makes the development of forensics software with advanced forensics techniques. According to the report (https://www.Statista.com/ Statistics/330695/Number-of-Smartphone-Users-Worldwide/ n.d.) published on August 20, 2020, about 1.37 billion smartphones were sold, and also it will reach higher in the next years. Smartphones like Apple iPhone, Samsung, Readme, and so on are much compact compared with computers, but the performance, storage, and features are getting advanced as technology is also increasing. Mobile devices have become an essential part of everyone's life, making criminal activities or being a part of a crime. No other computing devices like mobile devices handle a massive amount of sensitive information. Mobile devices hold an enormous repository of user data

FIGURE 4.2 Flow process of mobile forensics.

like event logs, messages, contact lists, credit, debit card numbers, memos, calendars, etc. These handheld devices are used to communicate with others worldwide, share photos and videos, be connected in social blogs, and much more. Since technology is developing, the mobile device becomes data carries as the network is available, they can keep track of all movements. With the rapid development of technology in mobile computing, criminals commit advanced crimes like hacking, malware attacks, phishing, and many more. In the majority of illicit drug peddler's cases, a mobile device has been used as a medium to contraband across borders. Major criminal organizations and terrorists use mobile devices to coordinate illegal activities and share information to commit the crime in a well-organized manner. Digital investigators can gain valuable insights from their mobile phones. Nowadays, social media networks like Facebook, Instagram, and many more users are selling drugs illegally. This information on mobile phones can help experts to find out the whole network.

Mobile Forensics is the sub-part of digital forensics used to recover the old events from mobile devices in a forensically sound manner with a proper flow process, as represented in Figure 4.2. Mobile device forensics is challenging during data analysis and recovery due to the increasing functions and similar functions like computers. However, the forensic examiner's main advantage is that they can recover the deleted data even after the guilt has tried to leave it unrecoverable.

The main aim of forensics investigators is to determine what, how, who, where, when, and why an incident happened. The incident response team or examiners should collect and analyze all relevant digital evidence to determine the guilt and all necessary questions. Each forensics software differs from the other. Some forensic software does not have write protection during the data acquisition. In contrast, some tools require removing the chip or installing a bootloader on mobile devices before extracting the information for forensics investigation. For all analyses, following proper guidelines and procedures is essential as mobile devices hold massive data and valuable data, which might help examiners. If the examiners failed to follow appropriate guidelines, it leads to the loss or damage of digital evidence.

4.3.1 Types of Data Acquisition

In every forensics examination, data acquisition is the primary task in recovering all details from providing the court's necessary evidence. There are about three types of data acquisition for mobile devices. They are Manual acquisition, Logical acquisition, and physical acquisition. During an investigation phase, evidence gathering is crucial for examiners to acquire relevant evidence without integrity. The manual acquisition method obtains the log files and copies all application files placed on the mobile devices. The logical acquisition is the process of bitwise copying all necessary information, which consists of a timestamp, location of data from the filesystem, and the physical acquisition acquires the data from memory, including deleted data. DMD, JTAG, and DD methods are performed by physical acquisition in Android devices. In contrast, sometimes, forensics tools like Encase, Cellebrite, OSF, Oxygen forensics, and much other software are used to perform logical acquisition. However, the iPhone's physical acquisition is acquired by the DD method, and iTunes logical acquisition can be achieved.

4.4   VARIOUS TYPES OF MOBILE EVIDENCE

The forensic analysis of the mobile devices varies based on the investigation type and the mobile device acquired from a crime scene. Evidence associated with mobile devices can be found in several locations, such as memory cards, SIM cards, and embedded memory. Not every investigation deals with analyzing all components associated with mobile devices. Each device is treated differently, and each forensics toolkit provides different reports based on its capabilities.

The baseline mobile phone provides standard information like IMEI (International Mobile Equipment Identity), User Contact list, Messages, call logs, memos, and notes on mobile devices. In contrast, smart mobile phones provide more information than baseline phones. The smartphone offers application installed information such as instant messaging apps, downloads, etc., and internet-enabled information like internet history, online purchase activities, etc. Hence, the source available in mobile devices is precious during an investigation. Initially, the researchers extracted the data for analysis and recovered the deleted information using forensics software. This session provides various types of mobile evidence that can pose valuable resources, as shown in Figure 4.3.

4.4.1   SMS/MMS

One of the potential pieces of evidence found in a mobile device is SMS and MMS. In traditional days, text messages were frequently used in baseline mobile phones, whereas today, the messages are communicated in social networks, also termed as instant messaging. However, many social network applications provide a platform for messaging, and it tends to ease of use for the user and becomes the default in smartphones. Text messages have the potential of providing full text, timestamp, date, and country code. Even it is possible to reconstruct the messages and contact lists sent and received in social media networks (Anglano, Canonico, and Guazzone 2017). Sometimes, the owner might have erased the messages from mobile devices, but the deleted messages can be recovered using data acquisition methods. Figure 4.4

FIGURE 4.3   Types of mobile evidence.

			SMS	Recharge	140	05-04-2017 21:58:05 (UTC+5:30)	R...	IN
✓	◎		SMS	VT-VFCARE	VT-VFCARE	05-04-2017 21:58:46 (UTC+5:30)	H...	IN
✓	◎		SMS	AT-650003	AT-650003	06-04-2017 09:02:58 (UTC+5:30)	S...	IN
✓	◎		SMS	BZ-SBIINB	BZ-SBIINB	06-04-2017 10:33:58 (UTC+5:30)	Y...	IN
✓	◎		SMS	BZ-SBIINB	BZ-SBIINB	06-04-2017 10:34:15 (UTC+5:30)	Y...	IN
✓	◎		SMS	AD-VPAYTM	AD-VPAYTM	06-04-2017 14:50:52 (UTC+5:30)	P...	IN
✓	◎		SMS	VM-155400	VM-155400	06-04-2017 17:49:55 (UTC+5:30)	M...	IN
✓	◎		SMS	AT-AIRMOV	AT-AIRMOV	06-04-2017 17:53:33 (UTC+5:30)	Y...	IN
✓	◎	🗑	SMS	VT-611112	VT-611112	04-01-2017 11:13:43 (UTC+5:30)	S...	IN
✓	◎	🗑	SMS	VK-UBERIN	VK-UBERIN	04-01-2017 14:23:01 (UTC+5:30)	D...	IN
✓	◎	🗑	SMS	VK-UBERIN	VK-UBERIN	04-01-2017 14:24:45 (UTC+5:30)	D...	IN
✓	◎	🗑	SMS	AT-650003	AT-650003	05-01-2017 10:29:11 (UTC+5:30)	A...	IN
✓	◎	🗑	SMS	AM-LIMERD	AM-LIMERD	05-01-2017 10:47:48 (UTC+5:30)	B...	IN

FIGURE 4.4 Text messages from Android mobile device.

represents the deleted and not deleted messages from a mobile device using the oxygen forensics toolkit.

4.4.2 CALL LOGS

Another key evidence for criminal investigation is analyzing the call logs from the mobile phone of the suspected. In Barmpatsalou et al. (2018), the researcher mentioned that most cyberbullying, cyber harassment, and illicit drug dealers prefer to communicate in calls or messages. This provides the user's mobile phone's behavioral pattern, which will help examiners have a clear insight into the suspect. There is much mobile forensics software available to retrieve the call logs, including name, number, timestamp, location, duration, country code, remote party, and mode of calls during a crime investigation, as shown in Figure 4.5.

4.4.3 MULTIMEDIA DATA

All mobile phones have diverse sources of evidence in audio and video files. Nowadays, smartphones have a wide variety of resources that allow users to save voice calls and

			Voice	Jadi Jio	+917904655421	22-03-2017 16:21:30 (UTC+5:30)	00:00:02	IN
			Voice	Jadi Jio	+917904655421	22-03-2017 15:57:23 (UTC+5:30)		IN
			Voice	Jadi Jio	+917904655421	22-03-2017 15:56:39 (UTC+5:30)		IN
			Voice	Indhu Airtel	+919952226656	22-03-2017 15:33:48 (UTC+5:30)	00:00:36	IN
			Voice	Jones	+919789576044	22-03-2017 15:26:49 (UTC+5:30)	00:00:13	IN
			Voice	Jones	+919789576044	22-03-2017 15:21:09 (UTC+5:30)	00:00:08	IN
			Voice	Jones	+919789576044	22-03-2017 15:15:53 (UTC+5:30)	00:00:31	IN
			Voice	Sundar Jio	+918072328407	22-03-2017 14:57:43 (UTC+5:30)	00:00:12	IN
			Voice	Jadi Jio	+917904655421	22-03-2017 14:17:40 (UTC+5:30)	00:00:18	IN
			Voice	Jones	+919789576044	22-03-2017 14:07:35 (UTC+5:30)	00:00:15	IN
			Voice	Sundar Jio	+918072328407	22-03-2017 13:29:15 (UTC+5:30)	00:00:29	IN
			Voice	Sundar Jio	+918072328407	22-03-2017 13:19:18 (UTC+5:30)	00:00:17	IN
			Voice	Sundar Jio	+918072328407	22-03-2017 12:42:57 (UTC+5:30)	00:00:28	IN
	🗑		Voice	Sundar Jio	+918072328407	22-03-2017 12:06:10 (UTC+5:30)	00:00:25	IN
	🗑		Voice	Sundar Jio	+918072328407	22-03-2017 08:56:19 (UTC+5:30)	00:00:09	IN
	🗑		Voice	Sundar Jio	+918072328407	22-03-2017 08:54:21 (UTC+5:30)	00:00:14	IN
	🗑		Voice	Jadi Jio	+917904655421	20-03-2017 20:09:04 (UTC+5:30)		IN
	🗑		Voice	Thenna	+918098234310	19-03-2017 12:01:08 (UTC+5:30)	00:00:12	IN
	🗑		Voice	Thenna	+918098234310	19-03-2017 12:00:40 (UTC+5:30)		IN
	🗑		Voice	+911408819040	+911408819040	19-03-2017 10:17:45 (UTC+5:30)		IN
	🗑		Voice	02230430101	02230430101	18-03-2017 21:28:30 (UTC+5:30)	00:03:52	IN
	🗑		Voice	Indhu Airtel	+919952226656	18-03-2017 19:58:29 (UTC+5:30)		IN
	🗑		Voice	Indhu Airtel	+919952226656	18-03-2017 19:58:22 (UTC+5:30)		IN

FIGURE 4.5 Retrieved Call Logs from a mobile device.

take videos with higher resolution. Sometimes, these audio and video files hold sufficient evidence in the courtroom to prove the victim's innocence. Audio evidence may include phone call logs, voice call recordings, including all the necessary information. There is a different session called video forensics for analyzing all kinds of video frame by frame. Through secret code, skype video recordings can be shared by a user. By analyzing the main(.db) database, the examiner can retrieve sent and received data in skype with public id, title, names, and timestamps (Yang et al. 2019).

4.4.4 Geolocation

The Geolocation Services by the user are analyzed, which helps the investigator to examine the location. Forensic experts can also find out when and where the suspect accessed the internet. Additional to this, network activities of the suspect device (Wi-Fi and GPS) can be analyzed. Figure 4.6 represents the web connections accessed on the mobile device.

The examiners can obtain the geolocation details in the GEO timeline tab where the suspect used the mobile device. With the help of inbuilt Maps and Routes, investigators can track the device owner's movement, which helps the investigator seize the criminals, as represented in Figure 4.7.

4.4.5 Browser History

A suspect mobile device's critical evidence can be found by analyzing web browser history, cache memory, cookies, and bookmarks. Due to their portability, ease of use, and mobility, mobile phones play a predominant role in everybody's life. In most criminal cases, a lot of digital evidence is stored on mobile phones. The researchers (Fang et al. 2012) extracted the user's web browser history to conduct a forensics investigation and retrieved all emails, browser searches, deleted history, etc. The web browser history from the mobile device is represented in Figure 4.8.

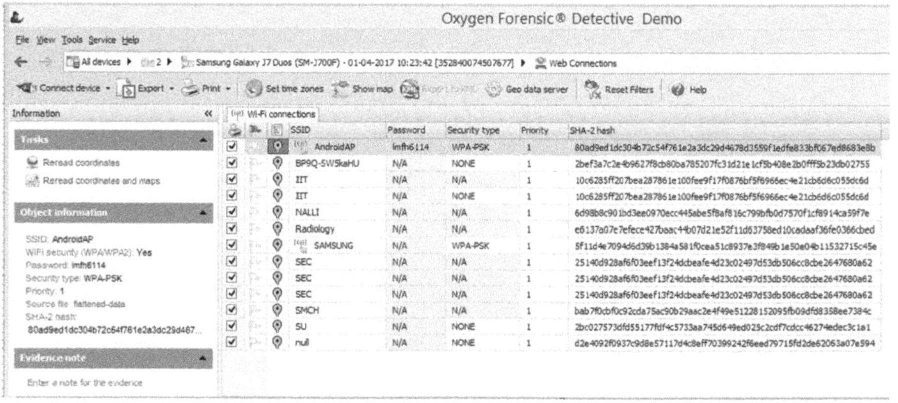

FIGURE 4.6 Details of Wi-Fi connected to a mobile device.

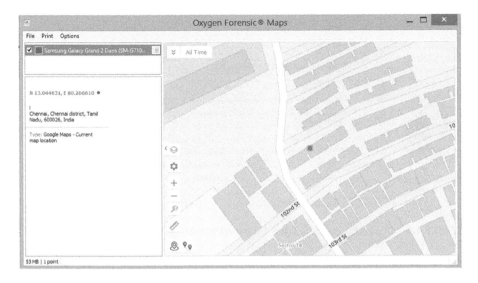

FIGURE 4.7   Location information from mobile devices.

FIGURE 4.8   Information from web browsing.

4.4.6 DEVICE APPLICATION

The device application section provides complete information about the mobile device and user applications installed in the user's mobile phone. It contains user social network applications, messages, address book, call logs, geo locations, visited places with coordinates and maps, deleted data, etc. The data from Google drive with attachments, last modified, last opened, modified by, and Google drive URL (uniform resource identifier) is also useful for the investigation purpose. If the criminal left any important information in emails or Google drive, it could easily be viewed by the examiners. Figure 4.9 represents the recovered login and password from mobile devices.

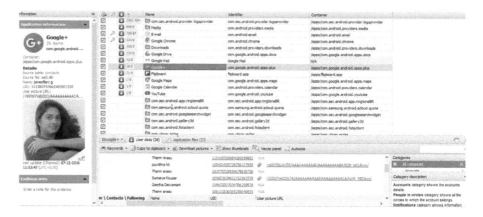

FIGURE 4.9   Login IDs, passwords are retrieved from this section.

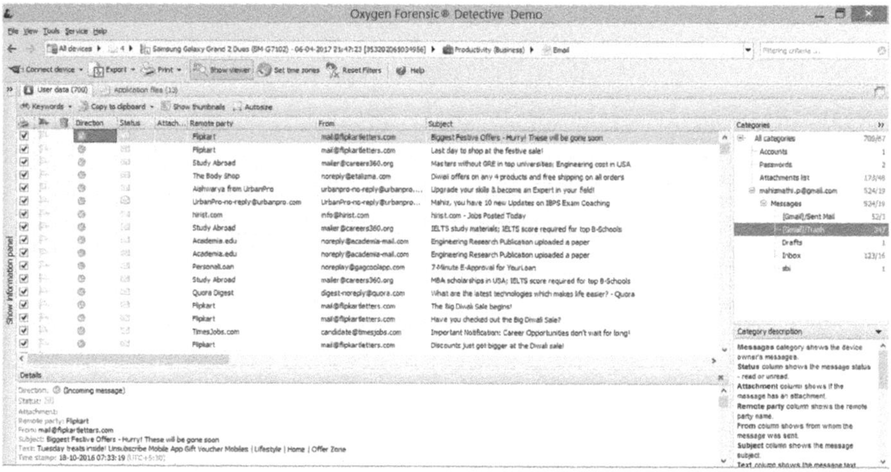

FIGURE 4.10   Email activities retrieved from a mobile device.

In today's world, email is also misused by criminals. An email has become the primary communication source, and almost everyone who owns a computer sends or receives emails daily. The fraud in emails has also increased day by day. The recovered email activities are represented in Figure 4.10. Some of the email frauds are email spoofing, threatening emails, phishing emails, etc. Google plus application allows the investigator to access the user details with names, user picture URL, and last update.

4.5   FORENSICS ACQUISITION AND EXAMINATION

The general workflow procedure is illustrated in Figure 4.1. Initially, the examiner has to check whether the concerned device is on or off. When the device is switched off, the bypass process will occur; if not, the data acquisition process is either done by forensics software or technique. Finally, the oxygen forensics technique is used for acquisition. The data acquisition is made with SD card and Android Backup data.

4.5.1 CREATING SOCIAL CONTEXT

The social network in a mobile phone constructs user friends, social community friends from WhatsApp, Facebook, LinkedIn, etc. In our design, the social graph is constructed where nodes and edges are formed from social networks. Figure 4.11 shows the call logs network of the user for the past two years. The node could have many properties like timestamps, country code, call duration, and SHA 2 hash value, while the edges measure the closeness with the user. Other properties include the communication status of an email, SMS, call, and Gmail. The histogram and matrix representation reveals the most active periods of device usage. It is built on timeline data collected from social networks. Figure 4.11 details the holder's social networking and contacts from the holder's phone and analyzes several call logs, messages, and social application activities. It is easy for examiners to use graph representation to identify social links to find common contacts and analyze communication statistics. It explores social network connections between the device holder and contacts. A user's social graph is constructed when the user registers with the system for the initial time. The construction process merges every time whenever the new user networks are available. The social network context graph is then stored in the heterogeneous data format, containing the data for efficient storage. In Figure 4.11, the node's size represents the user's volume of calls received or done. The geolocation gets stored in /apps/com.google.android. apps.maps.

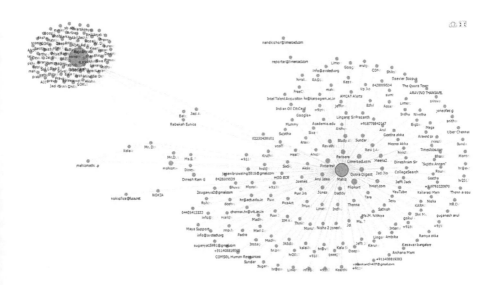

FIGURE 4.11   Entity-relationship between the user and contact list.

4.5.2 Data Analysis of Call Logs, Chat Communication, and Emails

The data analysis in the timeline section is summarized in Figure 4.12. The timeline analysis includes all call events, messages, and emails. The representation can be visualized in a bar graph either as a year, month, day, hour, minute, or second. The logs.db contains every communication, call, and email log with all necessary data. Once we have extracted the communication media from social networking, we can analyze it, as the electronic device from a crime scene cannot handle it due to data integrity. The unauthorized user can efficiently perform call spoofing, SMS spoofing, bluesnarfing. So, it is appropriate for the digital investigator to use the faraday bag to prevent external threats. Hence, the evaluation process and analysis of evidence will be carried out without data integrity. Most of the data is usually stored in internal memory, SIM cards, and external cards. In the digital investigator's term, it can be summarized that all the digital devices contain data that leave some traces for evidence in all places.

Consider a transaction taking place at one counter. When a transaction is completed, a system log is created. This data is called system data. For every device, there will be records that will help the digital examiner to identify the suspect. An enormous amount of information is available on everyone's devices in today's scenario. The example for unstructured data includes feedback forms, comments, a retweet from Instagram, Facebook, Twitter, etc. After retrieving the communication chats from social networking, the word cloud technique is used to find out the frequently used words. The digital data from an electronic device is unstructured data. One of the text mining approaches is integrating and querying text data after it is properly stored (Satiko et al. 2015). R-project is used as a visual representation, version 3.5.1. Some packages like ggplots, tm for text mining, word cloud, color brewer, and snowball are installed. The initial step is performed by analyzing the text from a social network application, then extracting the group chat from a social account using an oxygen forensics tool and generating word clouds at different levels. It is a visual

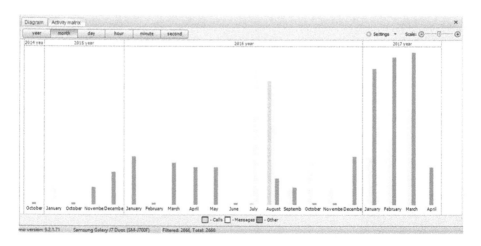

FIGURE 4.12 Timeline summarization of calls, messages, and others.

representation of frequently used words during specific group communication/chats, which are gathered in the form of text files.

There are about five steps to create a word cloud and find the frequently used words to identify the guilty. The following are the steps that are included,

- Extract the chats from smartphones that are found to be used by the suspect.
- The second is to transform the text file into a corpus, converting the text file into R readable format. It is a package to store the text document.
- The third process is to perform data processing. This step has to remove the whitespace, unwanted number, punctuation, etc.
- The fourth step is to convert the unstructured corpus format to a structured format. Document Term Matrix command is used to convert it.
- The final step is to make a word cloud to find out the frequently used keywords.

After extracting the chats from a group, it is transferred into a readable format and imported to the R program. Figure 4.13 represents the amount of frequently used crime and drug keywords in that group. From this, it is inferred if keywords related to the crime scene are found, and the results can be acquired. A word cloud can be created in three forms. They are a simple word cloud, a comparison cloud, a commonality cloud with tm (text mining), and word cloud packages. A word cloud gives several frequency words from a text file document by representing the different sizes of words in visualization. Generally, it is mainly used for quick survey responses, tweets, Facebook commands, or website content. It is also known as a social networking graph, which is used to predict the interaction between friends. Figure 4.13 shows that a word cloud can be created from group chats representing the visual presentation where huge words represent the more frequently used ones than other words. Digital Forensics helps to identify criminals. Machine Learning techniques can also be used to trace crime with greater accuracy based on historical data.

After extracting text messages from a mobile phone, machine learning algorithms are used to analyze the pattern. Additionally, a sentimental analysis was applied over the collected text messages, and performance measures were analyzed by model

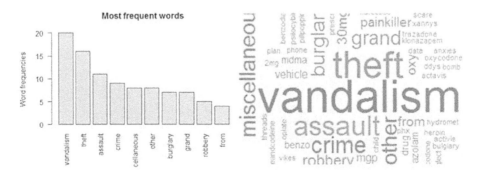

FIGURE 4.13   WordCloud and most frequency word. 4.6 machine learning and deep learning forensics.

accuracy and model loss for deep learning algorithms. Machine learning and deep learning algorithms are advanced techniques that can detect crime patterns (Jones and Winster 2020), detecting malware (Jones, Geoferla, and Winster 2020), and many more. Machine Learning algorithms were performed on forensically extracted data from mobile devices, whereas the deep learning algorithm was performed on Twitter data. The evidence image can also be analyzed using machine learning techniques using TensorFlow (Bhatt and Rughani 2017). This session provides insight into machine learning in mobile forensics data and can also help understand the user behavior pattern.

4.5.3 Pre-Processing

The process involved in pre-processing is termed the process of cleaning raw data before using machine learning algorithms. Text pre-processing ensures the removal of noisy data from the data set. Before encoding the text into a numeric vector, the pre-processing should be done by using the following techniques: Eliminating URL, converting all uppercase characters to lower case, eliminating unwanted characters like punctuation, removing stop words, stemming and lemmatization, Normalization, and many more techniques. After the pre-processing, each sentence from the dataset is categorized based on the sentence's sentiment. Figure 4.14 represents that the dataset has an equal number of positive and negative polarity.

4.5.4 Evaluation

The experimental analyses are done with four classification algorithms to classify the crime and drug keywords from the collected dataset. The algorithms are Random Forest, Gaussian Naïve Bayes, K-Neighbours classifier (KNN), and the decision tree classifier to classify the dataset. The author divided the dataset into 70% of training data and 30% testing data. In any classification problem in machine learning, the four

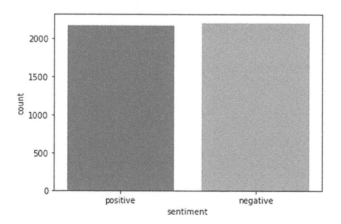

FIGURE 4.14   Polarity analysis for retrieved text messages.

TABLE 4.1
Confusion Matrix

$Accuracy(A) = (TP + TN_d/TP + FP + TN_d + FN_d)$	(4.1)
$Precision(P) = TP/TP + FP$	(4.2)
$Recall(R) = TP/TP + FN_d$	(4.3)
$F1score = 2PR/P + R$	(4.4)

combinations of predicted and actual values classification possibilities are used, as shown in Table 4.1. The performance metrics and confusion matrix are calculated in terms of accuracy, precision, f1 Score, and recall, as shown in Equations (4.1), (4.2), (4.3), and (4.4). Accuracy is defined as the ratio of the correctly classified patterns to the total number of patterns. The recall is termed as a ratio of correctly classified patterns to a total number of positive patterns. Precision is a ratio of truly classified patterns to the total amount of predicted positive patterns, and the F1*score is defined as the harmonic mean of precision and recall.

This experimental analysis investigates ML algorithms' applicability to identify and analyze illicit crime and drug evidence by analyzing historical data from mobile devices. After collecting the relevant data from smart devices through a forensically sound method, the data is pre-processed and analyzed with an ML supervised algorithm. The evaluation result presents SVM performs best with higher accuracy of 84% compared to NB, KNN, and Decision Tree with an accuracy of 80%, 79%, and 76%, respectively. Additionally, the ROC curve is also analyzed and represented in Figure 4.15.

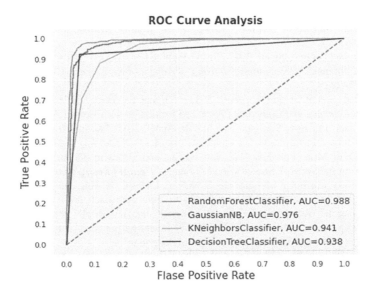

FIGURE 4.15 ROC curve.

4.6 CONCLUSION

The growing number of mobile devices is interconnected with the internet used for personal communication, which contains users, passwords, photographs, etc. Sometimes, these sources can act as potential digital evidence during an investigation. The most evident act is mobile devices will play a significant role in forensics examination and have a large role in a criminal investigation. Moreover, machine learning forensics is an advanced technique that helps the investigator find the user's behavioral pattern, malware analysis, detection pattern, and many more. This chapter presents the role of digital evidence available in mobile devices. The real-time data acquisition processes from mobile devices on various artifacts are also discussed with multiple potential resources. Another technology called machine learning and deep learning methods is incorporated with mobile forensics to understand a crime's pattern.

REFERENCES

Akbal, Erhan, Ibrahim Baloglu, Turker Tuncer, and Sengul Dogan. 2019. "Forensic Analysis of BiP Messenger on Android Smartphones." *Australian Journal of Forensic Sciences* 52 (5): 590–609.

Anglano, Cosimo. 2014. "Forensic Analysis of WhatsApp Messenger on Android Smartphones." *Digital Investigation* 11 (3): 201–213.

Anglano, Cosimo, Massimo Canonico, and Marco Guazzone. 2017. "Forensic Analysis of Telegram Messenger on Android Smartphones." *Digital Investigation* 23: 31–49.

Anglano, Cosimo, Massimo Canonico, and Android Smartphones. 2016. "Forensic Analysis of the ChatSecure Instant Messaging Application on Android Smartphones." *Digital Investigation: The International Journal of Digital Forensics & Incident Response* 19 (C): 44–59.

Barmpatsalou, Konstantia, Tiago Cruz, and Senior Member. 2018. "Mobile Forensic Data Analysis: Suspicious Pattern Detection in Mobile Evidence." *IEEE Access* 6: 59705–59727.

Bhatt, Prerak, and Parag Rughani. 2017. "Info Machine Learning Forensics: A New Branch of Digital Forensics." *International Journal of Advanced Research in Computer Science* 8 (8): 217–222.

Casey, Eoghan, and Benjamin Turnbull. 2011. "Digital Evidence on Mobile Devices." In *Digital Evidence and Computer Crime*, 3rd Edition, 1–44. Elsevier, Amsterdam

Denzel, Michael, Johannes Stüttgen, and Stefan Vömel. 2015. "Acquisition and Analysis of Compromised Firmware Using Memory Forensics." *Digital Investigation* 12: 50–60.

Divakaran, Dinil Mon, Fok Kar Wai, Ido Nevat, and Vrizlynn Thing. 2017. "Evidence Gathering for Network Security and Forensics." *Digital Investigation* 20: S56–S65.

Fang, Junbin, Zoe Lin Jiang, Mengfei He, S. M. Yiu, Lucas C. K. Hui, and K. P. Chow. 2012. The 7th International Workshop on Systematic Approaches to Investigating and Analysing the Web-Based Contents on Chinese Shanzhai Mobile Phones. https://www.researchgate.net/publication/288675257_Investigating_and_analyzing_the_web-based_contents_on_Chinese_shanzhai_mobile_phones.

Harichandran, Vikram, Frank Breitinger, Ibrahim Baggili, and Andrew Marrington. 2015. "A Cyber Forensics Needs Analysis Survey: Revisiting the Domain's Needs a Decade Later." *Computers & Security* 57: 1–13.

"https://www.statista.com/statistics/330695/number-of-smartphone-users-worldwide/"   n.d. Accessed November 17, 2020.

Jones, G. Maria, L. Ancy Geoferla, and S. Godfrey Winster. 2020. "A Heuristic Research on Detecting Suspicious Malware Pattern in Mobile Environment." *Test Engineering and Management* 3034: 3034–3041.

Jones, Maria, and Godfrey Winster. 2017. "Forensics Analysis on Smart Phones Using Mobile Forensics Tools." *International Journal of Computational Intelligence Research* 13 (8): 1859–1869.

Jones, Maria, and Godfrey Winster. 2020. "Analysis of Crime Report by Data Analytics Using Python." *Challenges and Applications of Data Analytics in Social Perspectives*, 54–79. IGI Global, London.

Koch, Robert, Mario Golling, Lars Stiemert, and Gabi Dreo Rodosek. 2015. "Using Geolocation for the Strategic Preincident Preparation of an IT Forensics Analysis." *IEEE Systems Journal* 10 (4): 1338–1349.

Landwehr, Norman, and Kenneth Landwehr. n.d. "Bind Torture Kill: The BTK Investigation." *The Police Chief: The Professional Voice of Law Enforcement* 73 (12): 16–20.

Rocha, Anderson, Walter Scheirer, Christopher Forstall, Thiago Cavalcante, Antonio Theophilo, Bingyu Shen, Ariadne Carvalho, and Efstathios Stamatatos. 2017. "Authorship Attribution for Social Media Forensics." *IEEE Transactions on Information Forensics and Security* 12 (1): 5–33.

Saleem, Shahzad, Oliver Popov, and Ibrahim Baggili. 2016. "A Method and a Case Study for the Selection of the Best Available Tool for Mobile Device Forensics Using Decision Analysis." *Digital Investigation* 16: S55–S64.

Satiko, Célia, Diego Miro, Carlos Francisco, and Simões Gomes. 2015. "Text Mining Business Intelligence: A Small Sample of What Words Can Say." *Procedia Computer Science* 55 (July): 261–67.

Tamma, Rohit, Oleg Skulkin, Heather Mahalik, and Satish Bommisetty. n.d. *Practical Mobile Forensics*, 3rd Edition. Packt Publishing, Birmingham.

Wu, Songyang, Yong Zhang, Xupeng Wang, Xiong, and Lin Du. 2017. "Forensic Analysis of WeChat on Android Smartphones." *Digital Investigation* 21: 3–10.

Yang, Teing Yee, Ali Dehghantanha, Kim-kwang Raymond Choo, and Zaiton Muda. 2019. "Windows Instant Messaging App Forensics: Facebook and Skype as Case Studies." *PloS One* 11 (3): e0150300.

Zickuhr, Kathryn. 2010. *Generations 2010*, pp. 1–29. Pew Research Center, Washington, DC.

5 Analysis of Kernel Vulnerabilities Using Machine Learning

Supriya Raheja, Rakesh Garg, and Bhavya Gururani

Amity University, Noida, India

CONTENTS

DOI: 10.1201/9781003140023-5

5.1 INTRODUCTION

The kernel is the pivot element of an operating system of any computing machine. It is basically a software program which controls the entire system of any computing machine. It is "the portion of the operating system code that is always resident in memory." Interaction between hardware and software is carried out and supported via kernel. Most of the systems, after loading Bootloader, load kernel as one of the initial programs on startups. So, the rest of the startup, along with every input and output requested by other software, translation of those requests into "data-processing instructions for central processing unit (CPU)" is carried out by kernel itself.

It manages not only memory but also other computer-linked devices like monitors, keyboards, speakers, printers, etc. To prevent any other "less critical parts of the operating system" from accessing the kernel, all its important codes are loaded and saved into a confided and safe space in the memory of the system. All the other "application programs" (such as word processors, web browsers, several video players as well as audio players), on the other hand, use "user space," which is a separate area of the system memory, as illustrates in Figure 5.1. In the confided memory space of the kernel, all the tasks it performs, such as managing interrupts, handling hardware devices, executing processes, etc., are protected.

This separation of user memory from the section of the memory containing kernel code ensures prevention of interference among user data and kernel data, thus preventing issues such as a malfunctioning application crashing the whole operating system or occurrence of slow processing and instability in the operating system.

"The kernel's interface is a low-level abstraction layer." The system call is the term addressed for the situation of the kernel being requested by any process. The design or kernel is separate in accordance with their way of managing these "system calls." The design is majorly divided into two types:

- A "monolithic kernel" is the kernel in which all the operating system's instructions run in the "same address space for speed."
- A "microkernel" is the kernel which runs almost all its processes in "user space, for modularity."

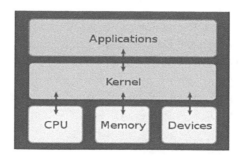

FIGURE 5.1   Interaction of kernel with other elements of a computing machine.

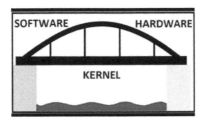

FIGURE 5.2   Kernel as a bridge between software and hardware.

The kernel has the responsibility of prioritizing and making a decision at any given time, which is the memory slot that each process can exploit and plan what the next course of action shall be in case of the absence of adequate memory space.

A kernel is neither software nor hardware but the bridge between the two of them, as shown in Figure 5.2. Kernel works behind the scenes and can be viewed only in the form of a "text log" printed by it. Otherwise, a computer user can never access the kernel directly. The kernel is basically "The heart of the operating system."

5.1.1 TYPES OF KERNELS

Kernels are of five types:

- "Micro Kernels"; consisting of base-level functionality
- "Monolithic Kernel"; consists of several "device drivers"
- "Exo Kernel"
- "Hybrid Kernel"
- "Nano Kernel"

A monolithic kernel is a very commonly used kernel by operating systems. For example, in Linux, "specifically loadable kernel modules," which are basically "device drivers," are often a part of a kernel. In case a device is required, enlargement of the kernel takes place by loading and joining the extension of the device. A fault in even a single one of the linked drivers, for instance, if one of the downloaded drivers is a "Beta driver," then the monolithic kernel can be troublesome and problematic because being a part of the kernel, this malfunctioning driver can override the particular process which is supposed to handle faulty programs. This can hamper the kernel from functioning, and by extension, the whole system would be ceasing to work. In case too many drives are present, the computer might turn slow or, in the worse case, would crash, as the kernel would run out of memory space.

Microkernels solve all the problems inflicted on the system by the monolithic kernel (Figure 5.3). Microkernels get involved in only the critically important activities like managing memory and CPU, in the case of a microkernel OS. All the drivers would be moved outside the kernel area in this case. Any malfunction in the case of a drive, for instance, a beta drive, would not crash the system. Instead, the kernel would just restart the system. Microkernel-based operating systems (Figure 5.4) are not common as, unfortunately, they are not easy to create. "Minix" and "QNX" are a few examples of existing Microkernels.

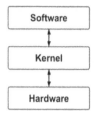

FIGURE 5.3   Monolithic kernel.

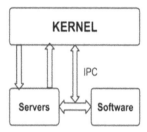

FIGURE 5.4   Microkernel.

5.2   COMMON VULNERABILITY EXPOSURE

CVE can be defined as a dictionary that stores all the information and defines vulnerabilities which have been publicly disclosed. CVE uses CVRF ("Common Vulnerability Reporting Framework"), which is under ICASI ("Industry consortium for advancement of security on the internet"). CVE can be termed as a dictionary which has definitions and information of publicly disclosed cyber security "vulnerabilities and exposure." The aim of CVE is to make the "sharing of data across separate vulnerability capabilities (tools, databases, and services)" easier with the help of the information CVE provides [1]. Some important "CVRF element" header information that defines CVE data are:

- **Document Title:** Document Title comprises of the date on which the data was last generated.
- **Date (Initial Release date, Current Release date):** Initial release date points out to the time the given document was produced. It follows the "UTC date/time" format. Although any kind of repository vulnerability database, there is not an initial release date.
- **Identification and Version:** "TIME STAMP" is used in the ID-element of CVE. CVE IDs are of the format "CVE-YYYY-NNNNN" where "YYYY" is the year that the vulnerability was introduced publicly and/or was assigned a CVE-ID. E.g., "CVE-1999-0067" is an ID number of vulnerabilities recognized publicly in the year 1999.

Finding and downloading more recently generated documents can be done using a "Simple string comparison routine." Version, on the other hand, is basically the process that involves the date and the time of the generation of the file to be encoded.

A CVE list comprises CVE entries. Each of these entries includes a CVE-ID ("CVE IDENTIFICATION NUMBER") with at least four digits in the "sequence number" portion of the IDE, information, and description, and at the least one public reference.

5.2.1 COMMON VULNERABILITY SCORING SYSTEM (CVSS)

CVSS is a platform open to all, used for discussing the "characteristics and severity of software vulnerability" [2]. There are three metric groups: "Base," "Temporal," and "Environmental."

5.2.2 BASE METRICS

The base metrics reflect the "intrinsic characteristics" of a vulnerability. There are properties or characteristics of the vulnerability that are perpetual throughout time in any kind of user environment. It has two sub-matrices:

- Exploitability metrics consisting of "Attack Vector, Attack Complexity, Privileges required, User interaction."
- Impact metrics consist of "Confidentiality impact, Integrity impact, Availability impact."

5.2.3 TEMPORAL METRICS

The Temporal metrics group consists of the properties of vulnerability that can vary over the period but remains constant through the user environment. This group contains the following metrics:

- Exploit Code Maturity
- Remediation Level
- Report Confidence

5.2.4 ENVIRONMENTAL METRICS

The Environmental metrics group consists of the properties of a vulnerability that are pertinent and distinct to the specific user environment. It takes into consideration the security control involvement, which may sabotage the attack. This group contains the following metrics:

- Modified Base Metrics
- Confidentiality Requirement
- Integrity Requirement
- Availability Requirement

5.3 BASE METRIC GROUP

The NVD (National Vulnerability Database) considers only the elements of base metrics, also known as a base metric group, for specifying Vulnerability [3, 4].

5.3.1 SCORING

An analyst assigns base value to a metric. This results in the derivation of a "base equation," which calculates a score ranging from "0.0" to "10.0". A vector string is produced by the "Scoring CVSS metrics," which is used in the scoring of vulnerabilities and represents the metric value in textual format. This vector string is a string that is always displayed with the vulnerability score and is specifically formatted in a way that has each value is assigned to each metric. A metric score is given under the assumption that the attacker has already identified and located all the vulnerabilities.

5.3.2 BASE METRICS VULNERABILITY COMPONENTS

These are the components of "base metrics" and are used to determine the vulnerability in the form of "scores."

- "Attack Vector"
- "Attack Complexity"
- "Privileges Required"
- "User Interaction"
- "Scope"
- "Confidentiality Impact"
- "Integrity Impact"
- "Availability Impact"

5.3.2.1 Exploitability Metrics

Exploitability metrics represent the "vulnerability component," i.e., the exact element causing the vulnerability. To derive the base equation, two sub-equations, namely "the exploitability sub-score equation" (Computed from the "base exploitability metrics") and "the impact sub-score equation" (Computed from the "base impact metrics") are used. When base metrics are scored, it is presumed that the attacker is well aware of the configuration of the software. All it is a default defense mechanism, for instance, rare limits, it is "built-in-firewall," etc., and all the other weaknesses [5].

5.3.2.1.1 Attack Vector (AV)

This metric depicts the factors that make vulnerability exploitation possible. The more remote an attacker is, both physically and logically from exploiting the components which are vulnerable, the higher is the value of the AV metric and, by extension, that of the Base Score. The assumption is that

> The number of potential attackers for a vulnerability that could be exploited from across a network is larger than the number of potential attackers that could exploit a vulnerability requiring physical access to a device, and therefore warrants a greater Base Score [6].

AV can take the following values

- Network (N): If a vulnerability has an AV spanning across the network, the attacker can exploit the vulnerability from any part of the vector. In this case, the vulnerability is said to be "Remotely exploitable." This value of the attribute also signifies that the vulnerability may be exploited using the internet's vastness. An example of this type of attack is CVE-2004-0230, where the network was attacked by a "denial-of-service" caused by an attacker. A TCP packet was specially created across a wide area.
- Physical (P): If AV has this value, it determines that the attacker must be in direct contact with the vulnerability-causing component in order to exploit it. "Cold Boot Attack" is an example of such an attack. In this case, the "Base Score" is lesser.
- Adjacent (A): This is similar to Network or "N" but makes a vulnerability comparatively difficult to be exploited by an attacker. Here the topology is such that the component that needs to be attacked must be adjacent to the one which is used by the attacker to exploit the vulnerability.

5.3.2.1.2 Attack Complexity (AC)

There are certain conditions existing that are not in the control of an attacker, which is required to be present so that vulnerability can be exploited. Attack Complexity metrics comprised of such conditions. These conditions may involve "computational exceptions" or may require more extraction of information regarding the target. Attack Complexity can be either high or low. As the value of attack complexity drops, the Base Score increases.

This attribute can take the following values:

- Low (L): This signifies that there is no specialized access or condition that is required by an attacker to gain access to the vulnerability. The rate of success of an attack is fairly high when attack complexity is low. This, in turn, causes "Base Score" to be greater.
- High (H): This establishes the fact that the complexity of conditions or access is high, and the attacker does not have any control over it. Though this does not make a vulnerability attack proof, it just makes it difficult for an attacker to successfully attack the vulnerability and exploit it. This causes "Base Score" to be lesser.

5.3.2.1.3 Privileges Required (PR)

There is a certain "level of privileges" an attacker must acquire before exploiting the vulnerability successfully. This metric includes the information and description of such privilege. These privileges can be either absent or vary from high to low. In the case of the absence of such privileges, the Base Score is the highest [7]. Following are the values that can be taken by this attribute:

- None (N): This determines that the attacker need not gain access to the "settings" or "files" before being able to reach the vulnerability. This causes the "Base Score" to be higher.
- Low (L): This is an intermediate value, which determines that the attacker needs to have some kind of access or "privilege" to be able to reach the vulnerability. Here the access might be easy, but the power over the software is lesser.
- High (H): Special "privileges" required to gain access to the vulnerable component of the software. Even though the access is not easily granted, the power is significant once the attacker gains access successfully.

5.3.2.1.4 User Interaction (UI)

This is the metric that depicts whether or not the given vulnerability requires the involvement of other human users (non-attacking) along with the attacker in order to successfully exploit the vulnerability, i.e., whether the attack is a "user-initiated process" or not. In cases of the requirement of a separate user, the Base Score is highest.

- None (N): This value determines that no user interaction is required. Henceforth it becomes easier for the attack to gain control and exploit the vulnerability successfully. It causes "Base Score" to be higher.
- Required (R): In order to be successful in exploring the vulnerability, the attacker needs to have a user first access the component. It causes "Base Score" to be lesser.

5.3.2.1.5 Scope (S)

Scope element determines if a software component which is "out of the Security Scope" of the vulnerable element is affected after the attacker successfully exploits the given vulnerability. A change in the scope after the attack results in an increase in the "Base Score."

5.3.3 Impact Metrics

Impact Metrics basically determines the extent of impact caused by different factors in software is the attacker is able to successfully exploit the vulnerability. It is important to know what components of the software are impacted and what would be the extent of the impact. Following are the attributes or metrics that are used to determine "Base Score" on the basis of impact [8, 9]:

- **Confidentiality (C)**

  As the name suggests, the value of the "Confidentiality" metric of the vulnerability is determined by how the confidentiality of the software and information affects if the attacker successfully deploys an attack and exploits the vulnerability. Authorization of confidential information is also taken into consideration. "Base Score" is higher with a lower impact.
  - High (H): Integrity is totally lost, and modifications can be made by the attacker easily. Difficult to access.
  - Low (L): Possibility of data modification is there but is indirect.
  - None (N): Integrity of the component is not lost.

- **Integrity (I)**
  The measure of loss of "Integrity" determines the values of the attributes. If an attacker gains access and is able to successfully exploit the vulnerability, "integrity" signifies how much of the component is affected in terms of its integrity with the software. "Base Score" is higher with higher "integrity."
  - High (H): High value signifies that there is a total loss in terms of integrity and protection loss. "Base Score" is higher.
  - Low (L): Modifications are possible by the attacker if access is gained, but there are some restrictions as well.
  - None (N): Integrity is not lost.
- **Availability (A)**
  This attribute is used to determine if, in case the component is exploited by the attacker, it remains available to the software or not. "Base Score" is higher is "Availability" impact is higher. This attribute can take the following values:
  - High (H): When the value is high, the software experiences a total loss of availability. Due to the complete inaccessibility of the component by the software, the attacker is able to gain full control over the component after successfully exploiting the vulnerability. "Base Score" is higher.
  - Low (N): In this, the component experiences a sudden reduction in performance or restrictions in resource access. Unlike when the value is high, the attacker is never able to gain total control over the component even after successful exploitation of the vulnerability. "Base Score" is comparatively lower.
  - None (N): When the impact value is "None," the availability of the component is not even slightly compromised if the attacker is able to successfully exploit the vulnerability. "Base Score" is lesser [10].

5.4   KERNEL VULNERABILITIES

Vulnerability in kernel depicts that the operating system, for example, Windows, Linux, Android, iOS, etc., have certain known vulnerabilities. Certain tools are available to scan out these vulnerabilities. These vulnerabilities have been taken under the account of CVE [2, 8, 11–15].

5.4.1   TOP FIVE LINUX VULNERABILITIES

The following vulnerabilities have been rated as "CVSS v2 10":

- *CVE-2017-18017*: This particular vulnerability was on the top of the Linux kernel vulnerability list of the year 2018 despite the fact that it is ID is dated to the year 2017 because it was first recognized and Got it is ID reserved in the year 2017 and later in January 2018 was published in NVD.
- *Linux kernel netfilter:* xt_TCPMSS description tells that the "tcpmss_mangle_packet" function in "net/netfilter/xt_TCPMSS.c" can allow remote hackers to carry out a denial-of-service attack.
- *CVE-2015-8812:* When the Linux kernel failed to define an error condition, a severe glitch was recognized in the Linux kernel's "drivers/infiniband/hw/

cxgb3/iwch_cm.c" of the Linux kernel." The outcome of this vulnerability was that "it allowed remote attackers to execute arbitrary code or cause a denial-of-service (use-after-free) via crafted packets."

- **CVE-2016-10229:** In this Linux vulnerability, the "udp.c" allows remote attackers to execute arbitrary code via UDP traffic that triggers an unsafe second checksum during the execution of a recv system call with the MSG_PEEK flag. The Google team discovered this vulnerability while using "the component for the Android mobile OS." It was noticed when a buffer smaller than the skb payload was provided.

- **CVE-2014-2523:** A serious glitch occurred in the "head for netfilter in the Linux kernel, this time by the incorrect use of a DCCP header pointer: Net/netfilter/nf_conntrak_proto_dccp.c". This vulnerability lets random attackers result in a system to crash or probably execute an alien code through a "DCCP packet that triggers a call to either the dccp_new, dccp_packet, or dccp_error function."

- **CVE-2016-10150:** This "use-after-free vulnerability" in the Linux kernel was discovered within the "virt/kvm/kvm_main.c's kvm_ioctl_create_device function". In this glitch of the operating system, the user is allowed to result in a "denial-of-service attack" or worse as it can allow the hacker to gain privileges via "crafted ioctl calls on teh/devkvm device" by simply exploiting this glitch.

These vulnerabilities can be scanned using several scanning methods. Some of these and other vulnerabilities have already been fixed as well. Other vulnerabilities with a score between 6–6.99 are shown in Figure 5.5.

#	CVE ID	CWE ID	# of Exploits	Vulnerability Type(s)	Publish Date	Update Date	Score	Gained Access Level	Access	Complexity	Authentication	Conf.	Integ.	Avail.
1	CVE-2019-12817	119		Overflow	2019-06-25	2019-06-28	6.9	None	Local	Medium	Not required	Complete	Complete	Complete

arch/powerpc/mm/mmu_context_book3s64.c in the Linux kernel before 5.1.15 for powerpc has a bug where unrelated processes may be able to read/write to one another's virtual memory under certain conditions via an mmap above 512 TB. Only a subset of powerpc systems are affected.

| 2 | CVE-2019-11599 | 362 | | DoS +Info | 2019-04-29 | 2019-05-28 | 6.9 | None | Local | Medium | Not required | Complete | Complete | Complete |

The coredump implementation in the Linux kernel before 5.0.10 does not use locking or other mechanisms to prevent vma layout or vma flags changes while it runs, which allows local users to obtain sensitive information, cause a denial of service, or possibly have unspecified other impact by triggering a race condition with mmget_not_zero or get_task_mm calls. This is related to fs/userfaultfd.c, mm/mmap.c, fs/proc/task_mmu.c, and drivers/infiniband/core/uverbs_main.c

| 3 | CVE-2019-11486 | 362 | | | 2019-04-23 | 2019-06-14 | 6.9 | None | Local | Medium | Not required | Complete | Complete | Complete |

The Siemens R3964 line discipline driver in drivers/tty/n_r3964.c in the Linux kernel before 5.0.8 has multiple race conditions.

| 4 | CVE-2019-6974 | 362 | | | 2019-02-15 | 2019-09-20 | 6.8 | None | Remote | Medium | Not required | Partial | Partial | Partial |

In the Linux kernel before 4.20.8, kvm_ioctl_create_device in virt/kvm/kvm_main.c mishandles reference counting because of a race condition, leading to a use-after-free.

| 5 | CVE-2019-3900 | 400 | | | 2019-04-25 | 2019-05-17 | 6.8 | None | Remote | Low | Single system | none | None | Complete |

An infinite loop issue was found in the vhost_net kernel module in Linux Kernel up to and including v5.1-rc6, while handling incoming packets in handle_rx(). It could occur if one end sends packets faster than the other end can process them. A guest user, maybe remote one, could use this flaw to stall the vhost_net kernel thread, resulting in a DoS scenario.

| 6 | CVE-2018-1000204 | 20 | | | 2018-06-26 | 2019-10-02 | 6.3 | None | Remote | Single system | Complete | None | None |

** DISPUTED ** Linux kernel version 3.18 to 4.16 incorrectly handles an SG_IO ioctl on /dev/sg0 with dxfer_direction=SG_DXFER_FROM_DEV and an empty 6-byte cmdp. This may lead to copying up to 1000 kernel heap pages to the userspace. This has been fixed upstream in https://github.com/torvalds/linux/commit/a45b599ad808c3c982f6cdc12b0b8611c2f92824 already. The problem has limited scope, as users don't usually have permissions to access SCSI devices. On the other hand, e.g. the Nero user manual suggests doing `chmod o+r+w /dev/sg*` to make the devices accessible. NOTE: third parties dispute the relevance of this report, noting that the requirement for an attacker to have both the CAP_SYS_ADMIN and CAP_SYS_RAWIO capabilities makes it "virtually impossible to exploit."

| 7 | CVE-2018-1000026 | 20 | | | 2018-02-09 | 2019-05-10 | 6.8 | None | Remote | Low | Single system | None | None | Complete |

Linux Linux kernel version at least v4.8 onwards, probably well before contains a Insufficient input validation vulnerability in bnx2x network card driver that can result in DoS: Network card firmware assertion takes card off-line. This attack appear to be exploitable via An attacker on a must pass a very large, specially crafted packet to the bnx2x card. This can be done from an untrusted guest VM.

| 8 | CVE-2018-18559 | 416 | | | 2018-10-22 | 2019-05-14 | 6.8 | None | Remote | Medium | Not required | Partial | Partial | Partial |

In the Linux kernel through 4.19, a use-after-free can occur due to a race condition between fanout_add from setsockopt and bind on an AF_PACKET socket. This issue exists because of the 15fe076e4ea787807a7cdc168df932544b58eba6 incomplete fix for a race condition. The code mishandles a certain multithreaded case involving a packet_do_bind unregister action followed by a packet_notifier register action. Later, packet_release operates on only one of the two applicable linked lists. The attacker can achieve Program Counter control.

| 9 | CVE-2018-16884 | 416 | | Mem. Corr. | 2018-12-18 | 2019-05-28 | 6.7 | None | Local Network | Low | Single system | Partial | Partial | Complete |

A flaw was found in the Linux kernel's NFS41+ subsystem. NFS41+ shares mounted in different network namespaces at the same time can make bc_svc_process() use wrong back-channel IDs and cause a use-after-free vulnerability. Thus a malicious container user can cause a host kernel memory corruption and a system panic. Due to the nature of the flaw, privilege escalation cannot be fully ruled out.

| 10 | CVE-2018-16880 | 787 | | Mem. Corr. | 2019-01-29 | 2019-05-16 | 6.9 | None | Local | Medium | Not required | Complete | Complete | Complete |

A flaw was found in the Linux kernel's handle_rx() function in the [vhost_net] driver. A malicious virtual guest, under specific conditions, can trigger an out-of-bounds write in a kmalloc-8 slab on a virtual host which

FIGURE 5.5 A snippet of vulnerabilities of Linux kernel with CVSS score b/w 6–6.99.

5.4.2 MICROSOFT WINDOWS KERNEL VULNERABILITIES

There are various software vulnerabilities that also exist in Microsoft software [16].

- **Buffer overflow**
  CVSSv3: 6.8 low
  CVE-ID: CVE-2019-1392
  CWE-ID: CWE-119 – "Improper Restriction of Operations within the Bounds of a Memory Buffer"
  Exploit Availability (A): No
  This glitch lets a user exceed the given privileges limit on the system, implicating an existence of "A boundary error when processing objects in memory within the Windows kernel." The hacker can basically "create a malicious application, launch it on the system and execute arbitrary code with system privileges."
- **Resource management error**
  CVSSv3: 5.1 low
  CVE-ID: CVE-2019-11135
  CWE-ID: CWE-399 – "Resource Management Errors"
  Exploit Availability (A): Yes ["Search exploit"]
  This window kernel vulnerability lets hackers access predictably "Sensitive information." The flaw can exist cause of a "boundary condition within the TSX Asynchronous Abort (TAA) in Intel CPUs."

Figure 5.6 represents the different vulnerabilities of Windows 10.

5.4.3 LIST OF SOME ANDROID KERNEL VULNERABILITIES

- "Sock_sendpage," CVE no.: CVE-2009-2693
- "KillingInTheNameOf psneuter ashmen," CVE no.: CVE-2011-1149
- "Exploid udev," CVE no.: CVE-2009-1185 [17]
- "levitator," CVE no.: CVE-2011-1350
- "Samsung GPU DMA"

Figure 5.7 shows the different types of Android kernel vulnerabilities.

5.4.4 TOP 10 "MOST COMMONLY EXPLOITED KERNEL VULNERABILITIES"

Following are the top ten commonly exploited kernel vulnerabilities among all operating systems:

1. "CVE-2018-8174 – Microsoft"
2. "CVE-2018-4878 – Adobe"
3. "CVE-2017-11882 – Microsoft"

#	CVE ID	CWE ID	# of Exploits	Vulnerability Type(s)	Publish Date	Update Date	Score	Gained Access Level	Access	Complexity	Authentication	Conf.	Integ.	Avail.
1	CVE-2019-1358	20		Bypass	2019-10-10	2019-10-15	2.1	None	Local	Low	Not required	Partial	None	None

A security feature bypass exists when Windows Secure Boot improperly restricts access to debugging functionality, aka 'Windows Secure Boot Security Feature Bypass Vulnerability'.

| 2 | CVE-2019-1359 | 119 | | Exec Code Overflow | 2019-10-10 | 2019-10-15 | 9.3 | None | Remote | Medium | Not required | Complete | Complete | Complete |

A remote code execution vulnerability exists when the Windows Jet Database Engine improperly handles objects in memory, aka 'Jet Database Engine Remote Code Execution Vulnerability'. This CVE ID is unique from CVE-2019-1358.

| 3 | CVE-2019-1358 | 119 | | Exec Code Overflow | 2019-10-10 | 2019-10-15 | 9.3 | None | Remote | Medium | Not required | Complete | Complete | Complete |

A remote code execution vulnerability exists when the Windows Jet Database Engine improperly handles objects in memory, aka 'Jet Database Engine Remote Code Execution Vulnerability'. This CVE ID is unique from CVE-2019-1359.

| 4 | CVE-2019-1347 | 119 | | DoS Overflow | 2019-10-10 | 2019-10-15 | 7.1 | None | Remote | Medium | Not required | None | None | Complete |

A denial of service vulnerability exists when Windows improperly handles objects in memory, aka 'Windows Denial of Service Vulnerability'. This CVE ID is unique from CVE-2019-1343, CVE-2019-1346.

| 5 | CVE-2019-1346 | 119 | | DoS Overflow | 2019-10-10 | 2019-10-15 | 7.1 | None | Remote | Medium | Not required | None | None | Complete |

A denial of service vulnerability exists when Windows improperly handles objects in memory, aka 'Windows Denial of Service Vulnerability'. This CVE ID is unique from CVE-2019-1343, CVE-2019-1347.

| 6 | CVE-2019-1345 | 200 | | +Info | 2019-10-10 | 2019-10-15 | 2.1 | None | Local | Low | Not required | Partial | None | None |

An information disclosure vulnerability exists when the Windows kernel improperly handles objects in memory, aka 'Windows Kernel Information Disclosure Vulnerability'. This CVE ID is unique from CVE-2019-1334.

| 7 | CVE-2019-1344 | 200 | | +Info | 2019-10-10 | 2019-10-15 | 2.1 | None | Local | Low | Not required | Partial | None | None |

An information disclosure vulnerability exists in the way that the Windows Code Integrity Module handles objects in memory, aka 'Windows Code Integrity Module Information Disclosure Vulnerability'.

| 8 | CVE-2019-1343 | 119 | | DoS Overflow | 2019-10-10 | 2019-10-15 | 7.1 | None | Remote | Medium | Not required | None | None | Complete |

A denial of service vulnerability exists when Windows improperly handles objects in memory, aka 'Windows Denial of Service Vulnerability'. This CVE ID is unique from CVE-2019-1346, CVE-2019-1347.

| 9 | CVE-2019-1342 | 20 | | | 2019-10-10 | 2019-10-15 | 7.2 | None | Local | Low | Not required | Complete | Complete | Complete |

An elevation of privilege vulnerability exists when Windows Error Reporting manager improperly handles a process crash, aka 'Windows Error Reporting Manager Elevation of Privilege Vulnerability'. This CVE ID is unique from CVE-2019-1315, CVE-2019-1339.

| 10 | CVE-2019-1341 | 269 | | | 2019-10-10 | 2019-10-15 | 7.2 | None | Local | Low | Not required | Complete | Complete | Complete |

An elevation of privilege vulnerability exists when umpo.dll of the Power Service, improperly handles a Registry Restore key function, aka 'Windows Power Service Elevation of Privilege Vulnerability'.

| 11 | CVE-2019-1340 | 269 | | | 2019-10-10 | 2019-10-15 | 7.2 | None | Local | Low | Not required | Complete | Complete | Complete |

An elevation of privilege vulnerability exists in Windows AppX Deployment Server that allows file creation in arbitrary locations. To exploit the vulnerability, an attacker would first have to log on to the system, aka 'Microsoft Windows Elevation of Privilege Vulnerability'. This CVE ID is unique from CVE-2019-1320, CVE-2019-1322.

| 12 | CVE-2019-1339 | 269 | | | 2019-10-10 | 2019-10-15 | 7.2 | None | Local | Low | Not required | Complete | Complete | Complete |

An elevation of privilege vulnerability exists when Windows Error Reporting manager improperly handles hard links, aka 'Windows Error Reporting Manager Elevation of Privilege Vulnerability'. This CVE ID is unique from CVE-2019-1315, CVE-2019-1342.

FIGURE 5.6   Snippet of Windows 10 kernel vulnerability.

#	CVE ID	CWE ID	# of Exploits	Vulnerability Type(s)	Publish Date	Update Date	Score	Gained Access Level	Access	Complexity	Authentication	Conf.	Integ.	Avail.
1	CVE-2019-9506	310			2019-08-14	2019-08-28	4.8	None	Local Network	Low	Not required	Partial	Partial	None

The Bluetooth BR/EDR specification up to and including version 5.1 permits sufficiently low encryption key length and does not prevent an attacker from influencing the key length negotiation. This allows practical brute-force attacks (aka "KNOB") that can decrypt traffic and inject arbitrary ciphertext without the victim noticing.

| 2 | CVE-2019-9461 | 200 | | +Info | 2019-09-06 | 2019-09-09 | 7.8 | None | Remote | Low | Not required | Complete | None | None |

In the Android kernel in VPN routing there is a possible information disclosure. This could lead to remote information disclosure by an adjacent network attacker with no additional execution privileges needed. User interaction is not needed for exploitation.

| 3 | CVE-2019-9458 | 362 | | | 2019-09-06 | 2019-09-09 | 4.4 | None | Local | Medium | Not required | Partial | Partial | Partial |

In the Android kernel in the video driver there is a use after free due to a race condition. This could lead to local escalation of privilege with no additional execution privileges needed. User interaction is not needed for exploitation.

| 4 | CVE-2019-9456 | 787 | | | 2019-09-06 | 2019-09-24 | 4.6 | None | Local | Low | Not required | Partial | Partial | Partial |

In the Android kernel in Pixel C USB monitor driver there is a possible OOB write due to a missing bounds check. This could lead to local escalation of privilege with System execution privileges needed. User interaction is not needed for exploitation.

| 5 | CVE-2019-9455 | 200 | | +Info | 2019-09-06 | 2019-09-09 | 7.1 | None | Local | Low | Not required | Partial | None | None |

In the Android kernel in the video driver there is a kernel pointer leak due to a WARN_ON statement. This could lead to local information disclosure with System execution privileges needed. User interaction is not needed for exploitation.

| 6 | CVE-2019-9454 | 787 | Mem. Corr. | | 2019-09-06 | 2019-09-09 | 4.6 | None | Local | Low | Not required | Partial | Partial | Partial |

In the Android kernel in I2c driver there is a possible out of bounds write due to memory corruption. This could lead to local escalation of privilege with System execution privileges needed. User interaction is not needed for exploitation.

| 7 | CVE-2019-9453 | 20 | | | 2019-09-06 | 2019-09-09 | 2.1 | None | Local | Low | Not required | Partial | None | None |

In the Android kernel in F2FS touch driver there is a possible out of bounds read due to improper input validation. This could lead to local information disclosure with system execution privileges needed. User interaction is not needed for exploitation.

| 8 | CVE-2019-9452 | 125 | | | 2019-09-06 | 2019-09-09 | 2.1 | None | Local | Low | Not required | Partial | None | None |

In the Android kernel in SEC_TS touch driver there is a possible out of bounds read due to a missing bounds check. This could lead to local information disclosure with System execution privileges needed. User interaction is not needed for exploitation.

| 9 | CVE-2019-9451 | 787 | | | 2019-09-06 | 2019-09-10 | 4.6 | None | Local | Low | Not required | Partial | Partial | Partial |

In the Android kernel in the touchscreen driver there is a possible out of bounds write due to a missing bounds check. This could lead to local escalation of privilege with System execution privileges needed. User interaction is not needed for exploitation.

| 10 | CVE-2019-9450 | 362 | Mem. Corr. | | 2019-09-06 | 2019-09-10 | 4.4 | None | Local | Medium | Not required | Partial | Partial | Partial |

In the Android kernel in the FingerTipS touchscreen driver there is a possible memory corruption due to a race condition. This could lead to local escalation of privilege with System execution privileges needed. User

FIGURE 5.7   Snippet of Android kernel vulnerabilities.

4. "CVE-2017-8750 – Microsoft"
5. "CVE-2017-0199 – Microsoft"
6. "CVE-2016-0189 – Microsoft"
7. "CVE-2017-8570 – Microsoft"
8. "CVE-2018-8373 – Microsoft"
9. "CVE-2012-0158 – Microsoft"
10. "CVE-2015-1805 – Google Android"

5.5   MACHINE LEARNING

Machine Learning is basically a "sub-area of artificial intelligence." The term infers to the property of IT systems finding the solutions to its problems independently by realizing and functioning on the reoccurring patterns of databases. In simple words: "Machine Learning enables IT systems to recognize patterns on the basis of existing algorithms and data sets and to develop adequate solution concepts." Hence, in ML, basic knowledge creates up artificial intelligence [10, 18–22].

People's action result is necessary to make it possible for the software to generate its own solutions independently. Let's consider for an instance that a necessary algorithm and involved data are loaded into the systems beforehand, and the set of rules for analysis that are foremost important in order to make the system recognize the patterns of the "data stock" is defined, then ML can come to act:

- "Finding, extracting, and summarizing relevant data."
- "Making predictions based on the analysis data."
- "Calculating probabilities for specific results."
- "Adapting to certain developments autonomously."
- "Optimizing processes based on recognized patterns."

5.5.1   TYPES OF MACHINE LEARNING

Algorithms are the most essential entity in machine learning. This section discusses the so the different types of machine learning algorithms.

- **Supervised learning:** In this case, the given input and output pairs are responsible for the system learning process. While the period of "monitored learning" would still be on, the programmer would feed the system with useful inputs. This would be like tutoring the system. The programmer must train the system in coherence with the "successive calculations," putting up connections with different sets of inputs and outputs.
- **Unsupervised learning:** In this case, AI ("Artificial Intelligence") doesn't get any "predefined target values" or any sort of "Rewards." It is functional for "learning segmentation (clustering)." The entered data is sorted according to their defining characteristics by the machine itself. For instance, if given coins to be sorted on the basis of different colors, a machine can easily learn and carry the task by differentiating in color.
- **Partially supervised learning**: This method involves both the "Supervised" and "Unsupervised" form of ML.
- **Encouraging learning:** This method works on the concept of "rewards and punishments." The algorithm is fed to the system by a "positive or negative interaction."
- **Active learning:** In this method, a system is allowed to work with and learn an algorithm in a way that the algorithm is allowed to "query results for specific input data on the basis of predefined questions that are considered significant" within a particular time frame. In general, the data basis can be either offline or

online, depending on the corresponding system. Mostly the algorithm can pick its own queries, whichsoever it considers to be of higher relevance.

This can be available either just once or recursively for machine learning. Another defining feature is the "either staggered development of the input and output pairs or their simultaneous presence." Hence, it creates a distinction between "sequential learning" and "batch learning."

5.5.2 RANDOM FOREST

The current work is using the supervised learning algorithm: Random Forest to analyze the vulnerabilities. A "Random Forests" (Figure 5.8) or "Random Decision Forests" are a complete self-suffced learning method for "classification," "regression," and similar other functions and tasks that perform by a unique way of "Constructing a multitude of decision trees at training time and outputting the class that is the mode of the classes or mean prediction of the individual trees" [9].

5.5.3 RANDOM FOREST REGRESSION

A Random Forest is a "meta estimator." Random forests regression works on the ideal concept of "Combination of predictions." Each "sub-sample size" here is handled by the parameter of "max_sample" if "boot_strap=true" (default). If not, the whole data is used up in the construction of a tree. The nature of a Random Forest regression can be denoted as given in Eq. (5.1).

$$g(x) = f0(x) + f1(x) + f2(x) + \cdots \qquad (5.1)$$

Where g is the "base model summation," "fi" is the "base classifier (decision tree)." Figure 5.9 shows the snippet for the Random Forest. Random Forest Regressor is implemented using the Scipy library in python. A loop is set to analyze the regressor for different values of "n estimator," and the appropriate value is chosen, and the "Regressor" is trained.

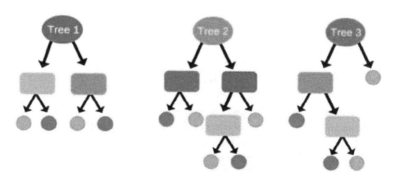

FIGURE 5.8   Random Forest ML algorithm.

```
>>> from sklearn.ensemble import Random
>>> from sklearn.datasets import make_r
>>> X, y = make_regression(n_features=4
...                           random_state
>>> regr = RandomForestRegressor(max_de
>>> regr.fit(X, y)
RandomForestRegressor(...)
>>> print(regr.predict([[0, 0, 0, 0]]))
[-8.32987858]
```

FIGURE 5.9 Snippet for the Random Forest ML algorithm.

5.6 METHODOLOGY ADOPTED AND DATA SET USED

The dataset for Linux kernel vulnerability is taken from the NVD database. The dataset used to train the Random Forest was curated out of two NVD CVE Feeds for the years 2019 and 2020. After processing, the data feeds are combined and then given to Random Forest Regressor for training, as shown in Figure 5.10. Original data feeds used from "https://nvd.nist.gov/" are mentioned in Table 5.1.

5.7 IMPLEMENTATION AND ANALYSIS RESULTS

Vulnerabilities logs are presented in Table 5.2. In Table 5.2, the independent entities can be classified as the "x-axis" of a corresponding graph, and dependent entities would be considered as "y-axis elements." It includes the attributes like availability, integrity, confidentiality, AV, score val, etc.

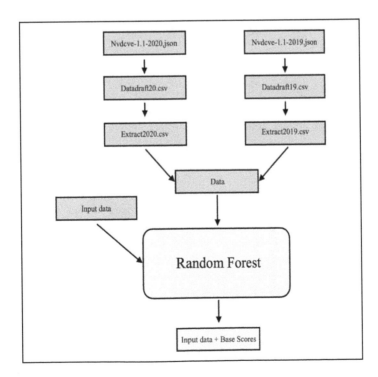

FIGURE 5.10 Methodology used.

TABLE 5.1
CVE Data Type and Data Format

	CVE_data_type	CVE_data_format	CVE_data_version	CVE_data_numberOfCVEs	CVE_data_timestamp	CVE_Items
0	CVE	MITRE	4	4404	2020-05-18T07:00Z	{'eve':{'data_type': 'CVE', 'data_format': 'M...
1	CVE	MITRE	4	4404	2020-05-18T07:00Z	{'eve':{'data_type': 'CVE', 'data_format': 'M...
2	CVE	MITRE	4	4404	2020-05-18T07:00Z	{'eve':{'data_type': 'CVE', 'data_format': 'M...
3	CVE	MITRE	4	4404	2020-05-18T07:00Z	{'eve':{'data_type': 'CVE', 'data_format': 'M...
4	CVE	MITRE	4	4404	2020-05-18T07:00Z	{'eve':{'data_type': 'CVE', 'data_format': 'M...
4399	CVE	MITRE	4	4404	2020-05-18T07:00Z	{'eve':{'data_type': 'CVE', 'data_format': 'M...
4400	CVE	MITRE	4	4404	2020-05-18T07:00Z	{'eve':{'data_type': 'CVE', 'data_format': 'M...
4401	CVE	MITRE	4	4404	2020-05-18T07:00Z	{'eve':{'data_type': 'CVE', 'data_format': 'M...
4402	CVE	MITRE	4	4404	2020-05-18T07:00Z	{'eve':{'data_type': 'CVE', 'data_format': 'M...
4403	CVE	MITRE	4	4404	2020-05-18T07:00Z	{'eve':{'data_type': 'CVE', 'data_format': 'M...

TABLE 5.2
Extracted Vulnerabilities Log

	Exploiyability Score	Impact Score	Attack VectorVal	Availability ImpactVal	Integrity ImpactVal	Scope Val	User InteractionVal	Attack CoplexityVal	Confidentiality ImpactVal	Priviledges RequiredVal	Base Score
0	2.2	3.6	4	3	1	1	1	2	1	1	5.9
1	3.9	5.9	4	3	3	1	1	3	2	1	9.8
2	2.2	3.6	4	3	1	1	1	2	1	1	5.9
3	1.8	3.6	2	1	1	1	1	3	2	3	5.5
4	3.9	1.4	4	1	2	1	1	3	1	1	5.3
5	3.9	5.9	4	3	3	1	1	3	2	1	9.8
6	3.9	6.0	4	3	3	2	1	3	2	1	10.0
7	3.9	5.9	4	3	3	1	1	3	2	1	9.8
8	1.8	3.6	2	3	1	1	1	3	1	3	5.5
9	3.9	3.6	4	3	1	1	1	3	1	1	7.5
10	2.8	3.6	3	3	1	1	1	3	1	1	6.5
11	2.2	3.6	4	3	1	1	1	2	1	1	5.9
12	3.9	3.6	4	3	1	1	1	3	1	1	7.5
13	3.9	3.6	4	3	1	1	1	3	1	1	7.5
14	2.8	2.5	4	1	2	1	1	3	3	3	5.4
15	2.8	3.6	4	3	3	1	1	3	1	3	6.5
16	2.8	5.9	4	1	3	1	1	3	2	3	8.8
17	2.3	2.7	4	3	2	2	2	3	3	3	5.4
18	3.9	3.6	4	3	1	1	1	3	1	1	7.5
19	3.9	5.9	4	1	3	1	1	3	1	1	9.8
20	1.8	3.6	2	3	1	1	1	3	2	3	5.5
21	3.9	5.9	4	1	3	1	1	3	2	1	9.8
22	2.3	2.7	4	3	2	2	2	3	2	3	5.4
23	2.3	2.7	4	1	2	2	2	3	3	3	5.4
24	2.3	2.7	4	1	2	2	2	3	3	3	5.4
25	2.3	2.7	4	1	2	2	2	3	3	3	5.4
26	2.3	2.7	4	1	2	2	2	3	3	3	5.4
27	3.9	3.6	4	3	1	1	1	3	1	1	7.5
28	1.8	5.9	2	3	3	1	1	3	2	3	7.8

```
ran=range(10,150,10)
plotd={'n_estimators':[],'model_score':[]}
for i in ran:
    X_train, X_test, y_train, y_test = train_test_split(X, y, test_size=0.2, random_state=0)
    regressor = RandomForestRegressor(n_estimators=i, random_state=0)
    regressor.fit(X_train, y_train)
    modelsc=regressor.score(X_test,y_test)
    plotd['n_estimators'].append(i)
    plotd['model_score'].append(modelsc)

maxsc=plotd['model_score'][0]
index=0
for i in range(len(plotd['model_score'])):
    if plotd['model_score'][i]>maxsc:
        maxsc=plotd['model_score'][i]
        index=i
print('MAX:\nn Estimator:',plotd['n_estimators'][index],'\nModel Score:',plotd['model_score'][index])

plt.figure(figsize=(15,8))
plt.xlabel('n Estimators')
plt.ylabel('Model Score')
plt.title('Model Tuning')
plt.plot(plotd['n_estimators'],plotd['model_score'])
plt.show()
```

FIGURE 5.11   Snippet to compute n estimator and model score.

In order to finely tune Random Forest Regressor and train the model effectively to have maximum "model score" or least errors, an appropriate value of "n Estimators" is to be decided. "n Estimators" defines the number of decision trees that are going to be used by the model to come to a final decision. Using the code illustrated in Figure 5.11, the value of "n Estimators" is determined.

The resultant maximum value is:

N Estimator: 120
Model Score:0.999934

The aim of this work is to determine "Base Score" as well as "Base Severity" for a software vulnerability using CVSS v3 metrics. This has been achieved, and the result is stored in "Result.csv." Following is a snippet of the resulting set of vulnerabilities with "Base Score" and "Base Severity" corresponding to it:

- The Base Score is the independent element and hence defines the x-axis.
- All the other elements are independent and belong to the y-axis.
- A Base Score is indicating severity.

The Base Score is going as high as 9.8 in the Linux vulnerability table, indicating the existence of several critical vulnerabilities (Table 5.3). The lowest value achieved is

TABLE 5.3
Base Score–Base Severity

Base Score	Base Severity
0	None
0.1–3.9	Low
4.0–6.9	Medium
7.0–8.9	High
9.0–10.0	Critical

4, which is of medium severity. Hence, we can conclude that the Base Severity is varying from medium to critical. The absence of vulnerabilities falling in "NONE" or "LOW" categories can be observed. It can be evaluated that in recent times, more harmful vulnerabilities have been detected.

5.8 CONCLUSION

The kernel is the major part of an operating system, as a link between hardware and software, and the unit that manages memory and other fronts of a system must be well protected in order to keep an individual user's data safe and system working. It is a necessity for a developer to keep a continuous and updated check on the kernel software to help keep these vulnerabilities at bay. The developer must relate to the various vulnerability databases that can be accessed from the internet. The database also requires to be updated by developers, which benefits the whole software developing community.

Using the metrics defined for each kernel vulnerability, the "Base Score" of a kernel software can be calculated, and hence the vulnerability can be easily analyzed. Through this work, it is achieved by using the Random Forest machine learning algorithm to calculate "Base Scores" and the "Base Severity." Base Severity helps us to analyze the general trend of the severity of vulnerabilities over a certain period of time and helps to keep track of the work that needs to be done in order to successfully patch these vulnerabilities. It is observed that the vulnerability from the recently updated vulnerability database is seen to be on the higher end of severity that shows that the vulnerabilities recorded recently pose more danger to the software.

Once a vulnerability is analyzed by its "Base Score" as well as "Base Severity," it becomes easy for the developer to plan the patch which is required to overcome the respective vulnerability.

REFERENCES

1. Cvedetails.com. 2021. CVE security vulnerability database. Security vulnerabilities, exploits, references and more. [online] Available at: https://www.cvedetails.com/vulnerability-list/ [Accessed 16 September 2021].
2. Christey, S., & Martin, R. A. (2007) Vulnerability type distributions in CVE. http://cve.mitre.org/docs/vuln-trends/vuln-trends.pdf.
3. Wijayasekara, D., Manic, M., Wright, J. L., & McQueen, M. (2012, June). Mining bug databases for unidentified software vulnerabilities. In *2012 5th International Conference on Human System Interactions* (pp. 89–96). IEEE, New York.
4. Zhang, S., Caragea, D., & Ou, X. (2011, August). An empirical study on using the national vulnerability database to predict software vulnerabilities. In *International Conference on Database and Expert Systems Applications* (pp. 217–231). Springer, Berlin.
5. Jimenez, M., Papadakis, M., & Le Traon, Y. (2016). Vulnerability prediction models: A case study on the Linux kernel. In *2016 IEEE 16th International Working Conference on Source Code Analysis and Manipulation (SCAM)* (pp. 1–10). IEEE, New York.
6. Mann, D. E., & Christey, S. M. (1999, January). Towards a common enumeration of vulnerabilities. In *2nd Workshop on Research with Security Vulnerability Databases*. Purdue University, West Lafayette, Indiana.

7. Krsul, I. V. (1998). *Software Vulnerability Analysis, Purdue University Ph. D* (Doctoral dissertation, dissertation, May).

8. Raheja, S., & Munjal, G. (2016). Analysis of Linux kernel vulnerabilities. *Indian Journal of Science and Technology*, 9, 48.

9. Ghaffarian, S. M., & Shahriari, H. R. (2017). Software vulnerability analysis and discovery using machine-learning and data-mining techniques: A survey. *ACM Computing Surveys (CSUR)*, *50*(4), 1–36.

10. Yang, M. (2020). *Using Machine Learning to Detect Software Vulnerabilities* (Doctoral dissertation, University of Toronto).

11. Chen, H., Mao, Y., Wang, X., Zhou, D., Zeldovich, N., & Kaashoek, M. F. (2011, July). Linux kernel vulnerabilities: State-of-the-art defenses and open problems. In *Proceedings of the Second Asia-Pacific Workshop on Systems* (pp. 1–5). Association for Computing Machinery, China.

12. Niu, S., Mo, J., Zhang, Z., & Lv, Z. (2014, May). Overview of Linux vulnerabilities. In *2nd International Conference on Soft Computing in Information Communication Technology*. Atlantis Press, China.

13. Mokhov, S. A., Laverdiere, M. A., & Benredjem, D. (2008). Taxonomy of Linux kernel vulnerability solutions. In *Innovative Techniques in Instruction Technology, E-learning, E-assessment, and Education* (pp. 485–493). Springer, Dordrecht.

14. Jimenez, M., Papadakis, M., & Le Traon, Y. (2016, December). An empirical analysis of vulnerabilities in OpenSSL and the Linux kernel. In *2016 23rd Asia-Pacific Software Engineering Conference (APSEC)* (pp. 105–112). IEEE, New York.

15. Breuer, P. T., & Pickin, S. (2006, June). One million (LOC) and counting: Static analysis for errors and vulnerabilities in the Linux kernel source code. In *International Conference on Reliable Software Technologies* (pp. 56–70). Springer, Berlin.

16. Raheja, S., & Munjal, G. (2021) Classification of Microsoft office vulnerabilities: a step ahead for secure software development. In Bhoi, A., Mallick, P., Liu, C.M., & Balas, V. (eds) *Bio-Inspired Neurocomputing. Studies in Computational Intelligence*, Vol 903. Springer, Singapore.

17. Shewale, H., Patil, S., Deshmukh, V., & Singh, P. (2014). Analysis of android vulnerabilities and modern exploitation techniques. *ICTACT Journal on Communication Technology*, 5(1), 863–867.

18. Witten, I., & Frank, E. (2005). *Data mining: Practical machine learning tools and techniques*. Elsevier, Amsterdam.

19. Han, J., Pei, J., & Kamber, M. (2011). *Data mining: concepts and techniques*. Elsevier, Amsterdam.

20. Owasp.org. 2021. Vulnerabilities I OWASP. [online] Available at: https://www.owasp.org/index.php/Category:Vulnerability [Accessed 16 September 2021].

21. Vaithiyanathan, V., Rajeswari, K., Tajane, K., & Pitale, R. (2013). Comparison of different classification techniques using different datasets. *International Journal of Advances in Engineering & Technology*, *6*(2), 764.

22. Arnold, J., Abbott, T., Daher, W., Price, G., Elhage, N., Thomas, G., & Kaseorg, A. (2009). Security impact ratings considered harmful. *arXiv preprint arXiv:0904.4058*.

6 Cyber Threat Exploitation and Growth during COVID-19 Times

Romil Rawat
Shri Vaishnav Vidyapeeth Vishwavidyalaya, Indore, India

Bhagwati Garg
Manager-Union Bank of India, Gwalior, India

Vinod Mahor
IPS College of Technology and Management, Gwalior, India

Mukesh Chouhan
Government Polytechnic College, Sheopur, India

Kiran Pachlasiya
NRI institute of Science and Technology, Bhopal, India

Shrikant Telang
Shri Vaishnav Vidyapeeth Vishwavidyalaya, Indore, India

CONTENTS

DOI: 10.1201/9781003140023-6

6.1   INTRODUCTION

The cybernated technologies they have relied on for a long time are their data center's, cloud web-mesh, departmental computers, and their now-virtual workers' digital devices used to remain linked to each other and the data of the enterprise -is even more important. The criteria imposed on digital web-mesh have skyrocketed overnight. COVID-19 themed phishing [1] scams began circulating in January, exploiting people's fear and uncertainty around the crisis. According to a security protection report, as many as 42% of endpoints are unprotected at any given moment, creating large numbers of potential vulnerabilities. With so many internet users working from home and using potentially compromised laptops or home computers to access their corporate web-mesh, these endpoints present a significant weak link in the surveillance chain. This is a key point: more workers working from home means a much larger risk surface and exponentially more endpoints to try and protect. Financial services companies are proving to be prime targets for cybercriminals during the COVID-19 crisis [2]. Some organizations reported receiving emails that appeared to be from the Global Healthy Forums and bore the signature of a fictitious doctor. The email had an attached document that claimed to carry guideline precautions against COVID-19, but in fact, when clicked, activated a Trojan horse malware that had been tailored for infiltrating banks. With maintaining Web-Archetype consistency and expression, the notion of protection involves bounding the approaches of an intruder for poisoning the system using forged values injections.

6.2   A WEB-MESH HOST THAT IS TRUSTED

1. A frank-but- inquisitive Web-mesh Host (NH).
2. A compromised/benevolent NH or a malicious NH.
3. Collusionary NH.

In all architectures, a benevolent/compromised NH [3] will interrupt all sorts of relay or inject false exposure alerts. Similarly, to execute user de-anonymization, a colluding NH may liaise with other malicious organizations. The NH is considered trusted within the unified architecture. It is liable for saving the encryption keys used to encrypt/decrypt TempIDs for users and handling them. If the system is hacked, this raises the possibility of data-stealing, a general threat to any centralized NH. The NH program needs to operate in a trusted abode in this sense and use sufficient protocols for authentication and access control. It is important to approve and protect all particulars shared between the NH and the Smartphone [4] of the customer and between the NH and healthy officials. Therefore, unified systems only consider malicious users in their ambush archetype and seek to keep all users' information protected in order to minimize the lack of privacy for users. This guarantees that no unauthorized third party is able to access or penetrate information from any information sent/ received. The archetype of ambush finds the government and the registry to be untrustworthy and only exposes to the healthy regulatory the identity of users. These architectures thus strive to mask user identity and create anonymous system IDs while escaping the NH's ability to connect user information to IDs. The decentralized architecture assigns data storage to the smartphones of customers, making the solution more resilient, such as the central NH, against a single point of failure/ambush [5]. However, a minimal expression al central NH is always needed by the decentralized architecture. Thus, a significantly reduced number of NH -based threats would be vulnerable. In Decentralized Designs Anonymous IDs [6] is submitted to the cloud and then theoretically available for matching by other smartphones. Therefore, if the holy NH has access to any side-channel information, it would not be able to learn any PIIs or to connect anonymous IDs, or generate any social graphics [7]. There will be little effect in the event of a data breach, as the interaction risk identification and notification procedures on the system were carried out by the hybrid architecture. This removes any ambush on re-identification/de-anonymization, in addition, additional methods are offered by the hybrid architecture to mask user identity from the NH while allowing for unified relay matching. It suggests the development of ephemeral IDs on computers [8], comparable to decentralized architectures. The reasoning is that computers retain total control of their hidden identifiers, leaving them less vulnerable to NH breaches. Figure 6.1 shows the Graph representing an increase in visiting malware and phishing websites during COVID-19.

6.3 OUR CONTRIBUTIONS

1. In an anonymous, distributed world, these Flooding threats [9] ambush are not trivial to counteract. Picked phishing scams and multiple ambushes.
2. An overview of relay and replay ambush techniques with a privacy trade-off focused on location cells. We prove that without making technological improvements, fake touch events can be effectively mitigated.
3. An overview of trolling ambush, in which systems used by positively diagnosed people are used to allow others to falsely assume that they are similar to those individuals, to spot inconsistencies and without relying on a trustworthy platform module.

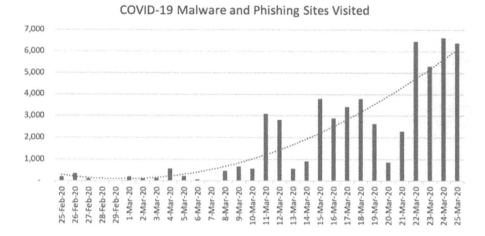

FIGURE 6.1 Graph representing an increase in visiting malware and phishing websites during COVID-19 [19].

6.4 RELATED WORK

Ahmed et al. [1] shown the first systematic probe of highly analyzed tracing entities. Also, offer an outline of several given tracing app scenarios, several implemented nationally, and address consumer questions about their application [10], outlining future research avenues for modern app architecture that promote better monitoring [11] and safety efficiency, as well as broad population acceptance.

Hawdon et al. [2] Live in periods unparalleled. The author has not, nor will, measure any fundamental explanation for why cyber victimization exists. The author cannot tell whether cyber victimization stays low as the social gap persists. However, the author may claim that the principle of repetitive tasks always holds. Live in extraordinary times, but the author's ideas somehow make sense of them.

Ventrella [3] A emphasis on rising cybercrime during the epidemic offers information on related vulnerabilities and solutions for personal data protection.

Yaron Gvili [4] provided a surveillance review of these requirements, the original version communicated to Apple Inc. early this April, shortly after the specifications were published. The author demonstrates that current standards can pose significant risks to society due to popular ambush and recommend new mitigation policies for risks not involving major technological changes and are simple to implement.

Khanday et al. [4] Extracted twitter data by application program interface, and annotation is done manually. Associated Hybrid feature innovation is conducted to pick the most important features. Using machine learning binary, binary tweet classification is performed. Decision tree results best than all other binaries. Develop feature engineering for improved performance and use deep learning for classification tasks.

6.6.6 SHINE A LIGHT ON SHADOW IT INFRASTRUCTURE

When they cannot be in the workplace together, the corporate team leaders and staff need opportunities to connect and cooperate. Some are going to press for digital solutions that you as a team have not accepted. Adopt such solutions. Encourage [22] the corporation to buy enterprise licenses for those solutions to be part of the debates on the acquisition and digital transformation. Help incorporate these methods, also ensuring that surveillance advice and safe settings are in place to help maintain controls of access, availability, and data loss protection. They will happen anyway, and if you do not, then you will have shadow IT problems.

6.6.7 ACCESS RESTRICTIONS ARE MORE RELEVANT THAN EVER BEFORE

Ideally, if you are not required to follow solutions, allow workers to differentiate their personal and job tasks using multiple devices [20]. For all privileged access, multi-factor authentication should be in effect. However, if key people are not present, you can also ensure that delegates are in place and "break glass" arrangements where possible. Do not just say that a single delegate is adequate.

6.6.8 KEEP UP THE CONTROLS ON DATA LOSS PROTECTION

Prevention of data loss aims to protect corporate IP and maintain the privacy standards of service and investigate alternatives for personal computer management if they need to be used for commercial purposes. Block or limit connections to removable storage players, monitors, and insecure home printers. Are workers really in need of USB media [15] connectivity at home?

6.6.9 KEEP THE STAFF AWARE OF RISKS

Through launching targeted phishing ambush, malicious actors and threat groups are leveraging the COVID-19 epidemic. These campaigns target the financial assets of staff or businesses. They try to solicit account credentials or unleash malware (including ransomware) onto enterprise web-mesh. As an important part of the COVID-19 relay plan [13], follow vulnerability alerts from credible threat information outlets to ensure that they are transmitted to your staff frequently. To help workers detect phishing scams. Ask workers and make it convenient for them, preferably, a single button on the email client) to report unwanted emails or files. With the surveillance world, express what you see; everybody is at risk.

6.6.10 BE ON GUARD FOR YOUR DEFENSE ACTIVITIES

Threat groups are taking advantage of the massive workload on IT and defense departments and are conducting malware threats, crypto-mining operations and denial of service ambush s at the business stage. The Surveillance Operations Center (SOC) and Disaster Relief Teams [10] cannot be used at a given time to operate or be able to operate virtual or with only a few participants on site.

Detection and quick reaction to cyber ambush means more now than ever. Ensure sufficient personnel and that when operating virtual, team members are well-practiced in handling threats. Provide them with a letter of authority stating their value to the enterprise if workers need to report to work and could be criticized by officials for doing so. To connect with and access data centers, rebuild web-mesh from physical copies, and failover/failback to resilience repositories, introduce and test alternative steps. Be sure you have backup relay systems for incident management teams and incidents that escalate. Most critically, ensure that you have deputies in organizational continuity and crisis response [12] departments for key employees in case team members become sick or are unable because of travel constraints.

6.6.11 TRACK THE CYBER HYGIENE OF THE WORKERS

Recognize that the workers are dealing in unfamiliar structures in unfamiliar ways. Encourage them to get support and stop a system of guilt if they are confused. To develop team spirit, line managers need to stay in touch with their teams and look out for staff that may feel lonely or maybe behaving in ways that pose questions. Ensure that workers often have opportunities to raise questions about work procedures, to help IT and surveillance [22] work safely promotes their roles, and to avoid "work-arounds," which can create safety problems.

6.6.12 CHECK THE PRIVILEGED USERS BY SANITY

Daily chats may help detect any fatigue or other conduct challenges that pose questions for workers with privileged enterprise access or system administrative privileges. Their well-being, team dynamics, and competitiveness may also be improved.

6.7 PROPOSED METHODOLOGY

The new COVID-19 had been in place, thoroughly tried and tested until the start of the epidemic, and we might have been well prepared for these proportions. If the planet faces similar or even much more serious situations, there is already a surge of development efforts around the world to create next-generation ambush detection software ready for immediate implementation. For the near term, we recognize the following as a significant point when evaluating research areas for potential ambush tracing applications.

6.7.1 IMPROVEMENT IN VICINITY ACCURACY

In our earlier debates, existing issues related to vicinity precision have been outlined. As a crucial part of the design process, the Bluetooth technology was not developed with position or vicinity determination. In addition, during the initiation process, new "Bluetooth-like" protocols [11] should be built with a focus on location/vicinity facilities. Actually, under their vicinity study, no COVID-19 tracing app has included such path capabilities. As hardware technology advances, the use of time-of-arrival measurements could also be used, and sub-nanosecond timing is commonplace within computers. Indeed, with Bluetooth vicinity research in mind, most Smartphone vendors have pushed to integrate a dedicated Ultra-Wideband (UWB) chipset. UWB uses 500MHz[17] relative to the low frequency used for Bluetooth (2MHz), theoretically resulting in cm-type sensitivity for vicinity determination. Research on integrating radio data with other positioning data from other phone-embedded devices promises to increase the precision of vicinity. Advances in technical solutions, such as advanced signal [15] processing for position monitoring and binary based on computational intelligence, may support these sensors [20]. New "vicinity" architectures directly aimed at incorporating all the related architecture into an isolated chip can be pictured.

6.7.2   DECENTRALIZED ARCHITECTONIC FOR INFECTION TRACING

One of the "takeaway messages" from the COVID-19 crisis is that the public needs to fix privacy issues for broader acceptance. None of the above-described applications [7] can be said to encompass a fully decentralized architecture – all of them use a central NH to different degrees, typically under the supervision of a governing authority. It is important to undertake research on a completely decentralized framework utilizing some form of peer-to-peer web-mesh to promote the exchange of privacy-preserving information among user devices.

6.7.3   DEEP LEARNING-BASED TECHNIQUES

Increasingly, AI binary is used as the computing capacity inside phones increases. The use is evident in helping the infection probability decision-making process. The binary will become "live" by metrics such as true infection recognition [19], missing detections, and false-positive performance, in the sense that they constantly adjust and enhance themselves in their reliability and accuracy. In this field, we expect a lot of studies, particularly with the incorporation of deep learning and decentralized architecture. Research effects are more difficult to forecast over a post-4.5-year time scale. A great deal depends on the advancement of hardware and the introduction of new innovations that are projected to become available. In the quantum arena, though, perhaps the most thrilling of these new developments reside.

6.7.4   QUANTUM COMPUTING

Many believe it to be on the verge of a breakthrough in both growth and commercialization. Future research on tracing solutions that maximize the usage of quantum computing's increasingly more efficient computing capabilities should start now. This may include quantum-based [18] deep learning binary and sophisticated tracking solutions of Monte-Carlo or particle philter form. In this paradigm, information from the phones would be sent for processing to a central quantum computer.

Another area of current research that is often touted to offer big developments in the coming years is quantum sensing. As a means of enhanced pacing, web-mesh synchronization, accelerometer accuracy, and position accuracy, to name a couple, this technology utilizes hypersensitivity found in quantum entanglement. Clearly, to improve the overall efficiency and effectiveness of the systems, all these problems should be taken to the touch tracing arena.

6.7.5 QUANTUM RELAY

Of all the recent quantum implementations, it is the most mature, at least in the context of implementation. Commercial offerings also exist in the field of a quantum relay, and there has already been proof-of-principle application in space. The main influence of this technology on the monitoring of apps is likely to be the advanced encryption [16] of relay and the increased protection of privacy it provides. In practice, in future virus tracking systems, surveillance and privacy will be absolute if correctly deployed, hacking and unauthorized access to virus tracking data will be made redundant.

6.7.6 PROBABILISTIC AMBUSHER

Profiles when adding Rocchetto's [3] ambusher profiles to an ambusher archetype, it is necessary to remember that one should not presume the ambusher will target a device in a real-world context. Two kinds of ambusher profiles, which are the static ambusher profile and the probabilistic ambusher profile, are used to archetype this non-deterministic behavior. Each of the six ambusher profiles was identified by Rocchettoet. Represents a static ambusher profile. [20]. Al. As a probability mass expression (PMF) of the six profiles, a probabilistic ambusher profile can be defined. The PMF is generated by assigning a probability of attacking (l_i) to each of the six ambusher profiles so that 0 alleged l_i ?? 1. The probability of attacking a particular ambusher profile is determined using:

$$P(\Delta_i) = \frac{l_i}{\sum_{j=1}^{n} l_j} \tag{6.1}$$

$\sum_{j=1}^{n} P(\Delta_j) = 1$ it is an example of a probabilistic ambusher profile designed to mimic the possibility of a nuclear power plant ambusher. At the outset of the ambush probe process, the probabilistic ambusher profile is sampled to achieve a covert ambusher profile that is known as the ambusher for the remainder of the attack process.

$$d_i = f(\Theta; \Gamma_i) = \sqrt{\sum_{j=1}^{n} \frac{1}{\beta_j^2} \left(\theta_j - \gamma_i^j\right)^2} \tag{6.2}$$

where β_j is a criticality factor such that $\{\beta \; ?? \; R|0 \; ?? \; \beta \; ?? \; 1\}$ increases the distance with a $\beta < 1$ criticality for properties. Every action's score (s_i) is inversely proportional to d_i and is determined using the following expression:

$$s_i = 1 - \frac{d_i}{\sum_{j=1}^{m} d_j} \quad i = 1,\dots,m \quad (6.3)$$

This equation is unique in that without adding nonlinear value-weighting, it measures the reciprocal of the distance as found in the inverse expression or exponential formulas such as the Softmax expression. The probability that the ambusher will take A_i action is determined using the expression, according to the score for each action:

$$P[A_i] = \frac{s_i}{\sum_{j=1}^{m} s_j} \quad (6.4)$$

Equation (6.4) has the intuitive understanding that the higher the score the ambusher receives for action, the greater the chance that the ambusher will choose this action.

6.8 DATA COLLECTION

Using the daily modified REST application programming interface (API) of Twitter [21], online discussions about the COVID-19 epidemic were gathered. We tried to delineate our probe to debate on the popular, country-specific hashtags relevant to the subject. Both data were processed in parallel for the comparative purposes of our study; that is, data sets were handled independently from each other, with comparisons drawn after the introduction of analytical strategies. Figure 6.3 shows the IoT attack dataset.

Before the advent of the '#BlackLivesMatter' movements, which have influenced national opinion globally, information was strategically time-bound for parsimony. We stored user schema, tweet schema, user engagement data, and hashtags information and URLs for each tweet used. Collectively, Twitter experiences task refers to the number of retweets, replies, mentions, and quotes. In order to catch the account for the propagation of some posts over others, retweets were included. The textual probe, hashtags were also preserved. Larger datasets of more global and dynamically modified search words were collected [24]. The final datasets consisted of larger datasets. For the purposes of this work, however, it was important to keep the geographically specific essence of the conversation constant; thus, for study, we retained this smaller but more technically acceptable dataset. It is important to remember that the words at the state level were not used. In order to identify a consumer as a bot, an 80% likelihood threshold was used for this table. Bot tweets refer to bots-generated tweets expected by the same threshold. For BotHunter forecasts and standard deviations for hate speech and offensive speech ratings, parentheses have dataset proportions.

2. Hawdon, J., Parti, K., & Dearden, T. E. (2020). Cybercrime in America Amid COVID-19: The Initial Results from a Natural Experiment. *Am J Crim Just* 45, 546–562.
3. Ventrella, E. (2020) Privacy in Emergency Circumstances: Data Protection and the COVID-19 Pandemic. *ERA Forum.* 21, 379–393.
4. Gvili, Y. (2020). Security Analysis of the COVID-19 Contact Tracing Specifications by Apple Inc. and Google Inc. *IACR Cryptol. ePrint Arch.* 2020, 428.
5. Khanday, A. M. U. D., Khan, Q. R., & Rabani, S. T. (2020). Identifying Propaganda from Online Social Networks During COVID-19 Using Machine Learning Techniques. *International Journal of Information Technology* 13(1), 115–122.
6. Ndiili, N. (2020). Unprecedented Economic Attack on Sub-Sahara African Economies: Coronavirus. *Environ Systdecis* 40, 244–251.
7. Uyheng, J., & Carley, K. M. (2020). Bots And Online Hate During The COVID-19 Pandemic: Case Studies in the United States and the Philippines. *Journal of Computational Social Science* 3, 445–468.
8. Siddiqui, M. F. (2021) Iomt Potential Impact in COVID-19: Combating a Pandemic with Innovation. In: Raza, K. (Eds) *Computational Intelligence Methods in COVID-19: Surveillance, Prevention, Prediction and Diagnosis. Studies in Computational Intelligence*, Vol 923. Springer, Singapore.
9. Deloglos, C., Elks, C., & Tantawy, A. (2020). An Attacker Modeling Framework for the Assessment of Cyber-Physical Systems Security. In: Casimiro, A., Ortmeier, F., Bitsch, F., Ferreira, P. (Eds) *Computer Safety, Reliability, and Security. SAFECOMP 2020. Lecture Notes in Computer Science*, Vol 12234. Springer, Cham.
10. Verma S., & Gazara R. K. (2021) Big Data Analytics for Understanding and Fighting COVID-19. In: Raza K. (Eds) *Computational Intelligence Methods in COVID-19: Surveillance, Prevention, Prediction and Diagnosis. Studies in Computational Intelligence*, Vol 923. Springer, Singapore.
11. Jnr, B. A., Nweke, L. O., & Al-Sharafi, M. A. (2020). Applying Software-Defined Networking to Support Telemedicine Health Consultation During and Post Covid-19 Era. *Health Technology* 11(2), 395–403.
12. Venkyanant, J. C., & Schwarz, A. (2020). *COVID-19 Crisis Shifts Cybersecurity Priorities and Budgets.* https://www.mckinsey.com/business-functions/risk-and-resilience/our-insights/covid-19-crisis-shifts-cybersecurity-priorities-and-budgets
13. Eian, I. C., Yong, L. K., Li, M. Y. X., Qi, Y. H., & Fatima, Z. (2020). Cyber Attacks in the Era of Covid-19 and Possible Solution Domains. Preprint.
14. Crisanto, J. C., & Prenio, J. (2020). *Financial Crime in Times of Covid-19 – AML and Cyber Resilience.* https://www.bis.org/fsi/fsibriefs7.pdf
15. Thierry Daumas, H., Wanklin, S., Van Der Linden, G., Kumar, S., Buvat, J., Subrahmanyam, K. V. J., Cherian, S., & Shahulnath, G. A., (2020). *Boosting Cybersecurity Immunity: Confronting Cybersecurity Risks in Today's Work-from Home World*, Capgemini Research Institute.
16. ECSO Recommendations on Cybersecurity in Light of the COVID-19 Crisis. Ecs-org. eu. 2021. https://www.ecs-org.eu/documents/uploads/ecso-recommendations-in-light-of-covid-19.pdf
17. Boismenu, P., Elbanna, M., Ben Rejeb, S., & Farrag, N. (2020) *COVID-19: Cyber Threat Analysis United Nations Office on Drugs & Crime Middle East And North Africa Assessment & Actions.* UNODC, Austria.
18. Venkyanant, J. C., & Schwarz, A. (2020). *COVID-19 Crisis Shifts Cybersecurity Priorities and Budgets, Cybersecurity Technology and Service Providers Are Shifting Priorities to Support Current Needs: Business Continuity, Remote Work, and Planning for Transition to the Next Norma.* Mckinsey and Company.

19. Duncan, A. (2020). Cybersecurity in the Covid-19 Era, An Infosys Consulting Perspective. https://www.infosysconsultinginsights.com/wp-content/uploads/2020/04/Cybersecurity-in-the-Covid-19-Era-POV.pdf
20. Raskar, R., Schunemann, I., Barbar, R., Vilcans, K., Gray, J., Vepakomma, P., Kapa, S., Nuzzo, A., Gupta, R., Berke, A., & Greenwood, D., (2020). Apps Gone Rogue: Maintaining Personal Privacy in an Epidemic. Preprint arXiv:2003.08567.
21. Modern Bank Heists 3.0. Vmware Carbon Black. www.Carbonblack.Com/Resource/Modern-Bank-Heists-3-0/
22. How Much Would a Data Breach Cost Your Business? IBM. www.ibm.com/Security/Data-Breach
23. Sjouwerman, S. (2020) Q1 2020 Coronavirus-Related Phishing Email Attacks Are Up 600%. Knowbe4. https://Blog.Knowbe4.Com/Q1-2020-Coronavirus-Related-Phishing-Email-Attacks-Are-Up-600
24. Storagecraft Research Reveals Rampant Data Growth, and Inadequate IT Infrastructures Are a Source of Global Concern and Risk. Storagecraft. www.Storagecraft.Com/Press-Releases/Storagecraft-Research-Reveals-Rampant-Data-Growth-And-Inadequate-It-Infrastructures

7 An Overview of the Cybersecurity in Smart Cities in the Modern Digital Age

Reinaldo Padilha França and
Ana Carolina Borges Monteiro

School of Electrical Engineering and Computing (FEEC) - State
University of Campinas (UNICAMP), São Paulo, Brazil

Rangel Arthur

Faculty of Technology (FT) - State University of Campinas
(UNICAMP), Limeira, São Paulo, Brazil

Yuzo Iano

School of Electrical Engineering and Computing (FEEC) - State
University of Campinas (UNICAMP), São Paulo, Brazil

CONTENTS

7.1 INTRODUCTION

Increasingly common worldwide, the smart city concept is transforming entire cities by using technology and intelligence in public management since smart cities are those that use connected devices to monitor and manage their businesses, streets, and public spaces. In its broadest sense, smart cities are urban centers that have been incorporating

IT technologies and solutions to integrate and optimize municipal operations, reducing costs and improving the quality of life of its inhabitants (Song et al. 2017).

It is evident, especially in large cities, that something must be done to improve the quality of life, public services, and sustainability. In addition to urban planning, it is necessary to invest in technological solutions that can be accepted and used by the residents of each smart city (Song et al. 2017; Cavada et al. 2017).

A city that reaches this level, therefore, is not only connected but a living and sustainable region that can use intelligence in favor of administration and resource management, as well as ensuring more safety and practicality in the use of roads and other devices public. One of the first things attacked in a smart city is related to one of the problems of any major urban center today: chaotic traffic. In this sense, systems integration acts as a catalyst for transformation. Where intelligent traffic lights are considered to receive satellite information, being able to automatically adjust timing to give more traffic senses (França et al. 2020h; Cavada et al. 2017).

With smartphone mobility and the support of IoT technologies, traffic agents can also work more efficiently and quickly by being directed to the most troublesome points or requesting signage maintenance within seconds. Another major point of concern in large urban centers is the safety of its inhabitants, where however efficient the police force is, it is often impossible to maintain the optimal proportion of agents to promptly respond to all occurrences. Soon in smart cities, more and more artificial intelligence monitoring is being used, considering security camera technology, security guards no longer need to be available 24 hours a day, as face recognition technologies can identify potential hazards and automatically trigger police (França et al. 2020h, 2020i; McClellan et al. 2017).

Sustainability is related to the bigger the city, the greater the concern with the management of resources, especially the natural ones, wherewith the implementation of technology in the public administration significantly increases the energy and water saving, besides enabling a most effective distribution to the inhabitants. A recent aspect used is the issue of entertainment that can be transformed through an integrated IT structure, where infrared sensors on the lampposts capture and record pedestrian shadows, which are projected by images to accompany who comes walking later (França et al. 2020h, 2020i; Visvizi and Lytras 2019).

Cities that already adopt sensor networks to prevent natural disasters, citizens who use smart keys to watch movies in cinemas and pay for subway tickets, and a city council connected to the social network. Or even pondering other initiatives around technology centers working with information to respond quickly to cases of traffic accidents and even expedite care for victims of landslides due to rain (Monteiro et al. 2020).

This all derives from the concept of smart cities, involving the adoption of massive technology to improve public services such as health, environment, safety, food, and transport, among others. However, it is essential that digital security is always in first place during the entire control of the systems, from the sensors to the processing of data in the cloud (Song et al. 2017; França et al. 2020h, 2020i; Monteiro et al. 2018).

This type of artistic insertion in the city's streets and squares improves the inhabitants' quality of life and encourages the better use of public space. With regard to tourism, this same kind of thinking can be explored with a focus on attracting people from other cities and countries, since information and tour guides integrated with smart city's system can be used in mobile apps to create custom roadmaps for each enriching the experience and driving the local economy (Coletta et al. 2018; Cavada et al. 2017).

Since most of the examples have some type of sensor for their monitoring, ranging from security cameras to rain control, the sensors are practically a fundamental part of a smart city, making up this set of sensors, the IoT technology (França et al. 2020a; Tran and Misra, 2019; Shovic 2016).

This chapter is motivated to provide a scientific contribution related to the discussion on the transformation in the governmental structure that smart city has generated, which is a complex and heterogeneous concept that involves the use of innovation based on ICT (Information and Communication Technology) to improve the quality of life in urban space. Since, in addition to being a process rather than a specific technical solution, it tends to be truly "a new way of governing" with a focus on sustainability and development. Also, discussing topics such as the aggregation of intelligence in microgrids, therefore, has a fundamental role in the proper manipulation of these energy resources along a grid, making these units contribute to a predetermined global good, along with the electric car that is associated with high technology by the automotive industry, and which has constantly been exhibiting as "the finest achievement of modern engineering" (França et al. 2020a; Choudrie et al. 2017).

However, these connected sensors have enormous potential for cybercriminals who exploit the highly flawed security of this, occurring due to the way that these devices are designed and operated, estimating that most of them are cheap and low maintenance, making them without updates security many times throughout its useful life (França et al. 2020a; Thames and Schaefer 2017).

Once collected by the sensors, the data needs to be processed in order to obtain intelligence, this processing of so much data is obtained through Big Data, which can optimize the development of cities, but even so, potential problems throughout this process can threaten security and privacy of residents (França et al. 2020a).

Thus, adopting the security of IoT-based communication networks in smart cities is also necessary to better understand the spread of malware, providing effective strategies for digital information protection. In addition, IoT devices, while offering very limited information processing capabilities, also require more efficient privacy protection mechanisms because it is very close to users. Data minimization, development of solutions to reduce the imminence and digital security flaws of data used and stored for the policy and management of the city, and access to information should be adopted, preventing the most sensitive information from being accessed without due permission (Arya et al. 2018; Yaqoob et al. 2017; Čolaković, Hadžialić 2018).

Therefore, this chapter aims to provide an updated overview of cybersecurity related to digital privacy and threats to personal data in a smart city context, showing its relationship with disruptive technologies, with a concise bibliographic background, featuring the potential of technology.

7.2   SMART CITIES CONCEPTS

Increasingly common in the world, the concept of the smart city is transforming entire cities by using technology and intelligence in public management since the idea involves sustainability, improvements in traffic, integration between public systems, energy, waste management, civil service offerings, and among other aspects. Related to the concept of "smart city" is to allow the evolution of the use of available resources, interpreting data and at the same time making that data become useful information for those in that city (Willis and Aurigi 2020, Mora and Deakin 2019).

However, it is important to emphasize that in a smart city, in which the citizen is the focus, it is complicated to develop all useful functionalities at once. In this sense, the most important thing is to think big but start small and quickly scale the results to achieve the goal. Since there are several aspects related to the use and administration of a city that can define a smart city, this concept goes far beyond the simplest and most direct way of considering that smart cities are those that use connected devices to monitor and manage the streets and public spaces (Nijholt 2019; Visvizi and Lytras 2019).

"Smart City" means that innovative urban space has been incorporating technologies ICT and IT solutions to integrate and optimize operations municipal, together with the efficiency of urban operations and services and competitiveness, meeting the needs of current and future generations in economic, social, and environmental aspects; it should be attractive to entrepreneurs, citizens, and workers, generating jobs and reducing inequalities, reducing costs and improving the quality of life of its inhabitants (Komninos 2019; França et al. 2020a).

It is valuing green spaces, optimizing electricity networks, and keeping greenhouse gas emissions low, in addition to concern with the proper use of natural resources, the elimination of garbage collection, and the improvement of traffic through the use of technology results in sustainable development serving as support for achieving a balance in the development of smart cities. However, environmental concern not only raises awareness about consumption but also seeks to reduce pollution and contamination of natural resources, as long as water and waste management, pollutant gas emission rates, CO_2 emissions, and energy consumption electricity appear more comprehensively in the evaluation of urban sustainability, encompassing actions that imply not only in the saving of operating expenses but in the reduction of everything that interferes negatively in the environment (Kamyab et al. 2020; Zawieska and Pieriegud 2018; Ramaswami et al. 2016).

Smart Cities are urban centers focusing on the fields of urban development related to mobility, safety and health, education, economy, environment, and government, which are the main axes that must be observed in a smart city (Komninos 2019; Nijholt 2019).

7.2.1 Technological Aspects Applicable to Smart Cities

A city that reaches this level, therefore, is not just connected, but a living and sustainable region that manages to use intelligence in favor of administration and resource management, in addition to ensuring more safety and practicality in the use of roads and other devices public. Through sustainability considering the green areas of the city, combined with the preservation of the environment in terms of the reduction in the consumption of fossil fuels and the use of renewable energy, the reuse of waste, and the permanent monitoring of air quality, which are essential characteristics of a smart city, has a positive impact on the economic aspect of electricity, for economic growth coupled with sustainable development. Each smart city has its specificities, but all have the common goal of providing its residents with a more fluid, cheap, sustainable, and intelligent relationship (Nijholt 2019; Visvizi and Lytras 2019; Willis and Aurigi 2020).

Traffic is one of the first aspects attacked in a smart city is also one of the biggest problems of any major urban center today: chaotic traffic. In this sense, systems integration works as a catalyst for transformation, since with intelligent traffic lights, it receives satellite information, being able to automatically adjust the timing to give fluidity to the directions with more traffic (Willis and Aurigi 2020).

And in this sense, with the mobility of smartphones and the support of IoT technologies, traffic agents can also work more efficiently and quickly by being directed to more problematic points or requesting the maintenance of signs in a matter of seconds (Mora and Deakin 2019; Farsi et al. 2020).

Security is another major point of concern in large urban centers, where, however efficient the police force may be, it is usually impossible to maintain the ideal proportion of agents to respond promptly to all occurrences, so in this sense, smart cities have been getting increasingly more in monitoring by artificial intelligence, through security cameras, the guards no longer need to be available 24 hours a day, since facial recognition technologies identify possible risks and automatically trigger the police (Visvizi and Lytras 2019; França et al. 2020h, 2020i).

Automated monitoring systems can be even more useful for personnel control not only to ensure security and to identify strangers within a certain location in the city, but the technology speeds up credential verification, making access and editing more reliable confidential information, in addition to providing relevant information on the use of space so that the entire internal operation of a given location can be redesigned (Willis and Aurigi 2020).

Entertainment is also an important point, which can be transformed through an integrated IT structure, in cities on the level of the smart city use infrared sensors on lampposts to record pedestrian shadows and project images to accompany them whoever walks afterward, as long as this type of artistic insertion in the streets and squares of the municipality improves the quality of life of the inhabitants and encourages the best use of public space (Mora and Deakin 2019).

With regard to tourism, it can be exploited to attract people from other cities or even from other countries, even considering that information and tourism guides integrated into the city system can be used in mobile applications creating personalized itineraries for each visitor, enriching the experience and driving the local economy (Al-Turjman 2020; Komninos 2019).

As in smart cities, automation in the management of these systems leads to significant savings, where technology can be used to control energy consumption, mainly by shutting down systems when in disuse, also identifying the biggest resource spenders, developing plans for readjustment and redesign of processes to spend less without affecting productivity (Nijholt 2019; Townsend 2013).

The philosophy of systems and processes integration performs automated traffic control as a reference for the management of a city, without productive bottlenecks and people trapped in a slow system, as well as the use of the IoT and mobile applications that agents and maintenance workers use to solve problems around the city so that all departments perform the most urgent duties immediately and are always where the company needs them to be (Townsend 2013).

The benefits brought by a smart city to the public agent are related to the reduction in the cost of sending letters to notify the citizen about the request made; reducing costs with the volume of paper stored; creating a dashboard to find out how many orders are requested, granted and rejected; agility in obtaining and disseminating information; the possibility of digitally tracking all stages of the process to improve performance; as well as making it possible to reduce the number of employees serving the public, allocating them to more critical tasks (Visvizi and Lytras 2019).

In the same sense that benefits are seen in relation to the citizen in the agility in obtaining information; the possibility of performing the procedure digitally, at any time or place; sending material inside the portal in a simple way; the single access point to interact with various services of the city hall; and the increase in the transparency of the process steps (Mora and Deakin 2019).

Transforming traditional cities into smart cities is a relevant demand for development in some cities focused on this concept, which should also be based on the international references identified. It is understood that cities transformed into smart cities should be used as benchmarks, not those already built on that concept. Since the idea of a smart city is to offer more well-being and quality of life to citizens through technology and the advantages it provides, therefore, there is a wide range of possibilities to start the smart city project, having with the objective of making the citizen able to carry out activities that in fact require his presence, as a result, generating savings for public coffers (Farsi et al. 2020).

7.3   THE IMPORTANCE OF CYBERSECURITY IN SMART CITIES

A smart city refers to the composition of a city that makes use of digital technologies to interconnect, preserve and improve the lives of the population, involving the adoption of massive technology to improve public services such as health, environment, security, food, and transport. This is done by sensors installed in the city in the following areas such as parking availability, lighting, waste management, traffic, and information about the bus. Collecting data through these sensors can lead to significant improvements in the way the city's infrastructure is used and a better understanding of urban issues (Conti et al. 2018).

Providing some examples of technologies used in addition to those already mentioned are intelligent and adaptable lighting according to need and demand, population monitoring through digital video, fire control management, and public announcement systems, intelligent roads with warnings, messages, and detours according to climatic conditions and unexpected events such as accidents or traffic jams, waste management with the detection of garbage levels in containers to optimize the garbage collection route, among many others (Thames and Schaefer 2017; Gupta 2018).

Most of these examples have some type of sensor for their monitoring since the sensors are practically a fundamental part of a smart city. However, these connected sensors have enormous potential for cybercriminals to be exploited, i.e., highly flawed security. This is due to the way that these devices are designed and operated, that is, most of them are inexpensive and low maintenance, making them without security updates many times throughout their useful life (Farahat et al. 2019).

The data is sent to a central that analyzes the information in real-time and gives the smart city authorities information that allows them to adjust the amount of energy it uses on the streets, the number of trucks needed for garbage collection during the week, and the volume of water used to irrigate city parks, for example, this reduces your electricity and garbage collection costs (Song et al. 2017).

However, since the benefits are countless, on the other hand, the smart city can collapse if cyber-attacks take place against its sensitive and critical digital infrastructures. Given the complexity of the digital infrastructure of smart cities, points of vulnerability are created, and the possible occurrence of a digital security incident will lead to essential services not functioning properly or being interrupted as water and energy supplies. Since the breach in cybersecurity opens loopholes for systems to be managed virtually, it can be subject to hacker attacks that compromise not only the privacy but the lives of millions of people who use its services (Alibasic et al. 2016; Bhardwaj et al. 2016a; Rawat and Ghafoor 2018).

However, all the benefits provided by the use of new technologies in cities can be wasted if due care is not taken with cybersecurity. A smart city makes use of technologies related to IoT, IIoT (Industrial Internet of Things), and IoE (Internet of Everything) for the functioning of services and applications. These technologies can become a threat because those who provide the solutions usually have no sense of good cybersecurity practices. Cybersecurity systems are rarely implemented and, when a vulnerability is found, the service hardly stops working to be updated, as it is vital for society (Alibasic et al. 2016; Jayapandian 2019).

Likewise, that unauthorized access to citizens' personal data can lead to major breaches of privacy. Assessing those smart cities have very vulnerable components that can provide cyber-attacks, such as the convergence between the physical and the cyber world, which will become practically the same, allowing vulnerabilities to be exploited on one side and take the devastating effects to the other using connections between them. Or, even due to the interoperability that exists between systems in smart cities, the surface of cyber-attacks becomes broader, making it necessary to increase cybersecurity and all its aspects involved. Or even with respect to the correlation between complex and large-scale processes in the services provided by smart cities, it can be an attraction for attacks with financial justification (extortion) (Bhardwaj et al. 2016b, 2016c; Rawat and Ghafoor 2018; Qu et al. 2019).

7.3.1 Security Challenges to Smart City Networks

Application-targeted cyber-attacks are those in which a smart city's digital platforms receive data from sensors and send commands to controlled devices, generally used in local and cloud applications. This is one of the main vectors for cyber-attacks, and its possible targets are collaborative digital platforms, portals of public administration and government agencies, portals of services available to citizens of the smart city, and applications in general (Bhardwaj and Goundar 2018; Ismagilova et al. 2020).

Or even mention the types of attacks on the integrity and confidentiality of data in relation to the storage and processing of the large volume of citizen data and smart city control applications that require databases with a high level of performance, integration, and analysis, which will later provide information intelligently.

The implementation of AI systems helps cybersecurity professionals to face attacks and protect the IoT data traffic network, and sensitive data is sent through the cloud systems, in which case the tools need to be implemented to detect and prevent the use of encryption to mask malicious activity. Since AI is able to automatically learn and detect unusual patterns in web traffic environments in encrypted form, that is, it will help network security defenses (Bhardwaj et al. 2020; Alshehri et al. 2021).

With AI, many cyber-attacks will be prevented from causing damage before it even reaches the end-user. This is because it is understood that AI is connected all the time, making it possible to learn about various vulnerabilities in real-time, doing nothing be new to your preventive actions. Another way in which AI systems can work is by categorizing digital cybersecurity attacks based on the level of threat, making systems more accurate enough, which can increase effectiveness over time, giving a dynamic advantage over cybercriminals (Beebe 2019; Bhardwaj and Goundar 2019).

The data protection policy, as well as the recommended actions for certain situations, can be applied in an automated way with the AI. This will not only save time but should also ensure that actions can be treated uniformly and thus have quicker solutions or deficiencies that are easier to correct. Another AI resource that can accompany these functionalities is forensic analysis, capable of not only learning from the user and making automated decisions, but also recording all actions performed on the network, resulting in a more precise human involvement (when necessary) in solving punctualities, i.e., automated decision making based on organizational compliance.

7.5 TRENDS

Passwords are due to the fact that each time they become easy to be broken in cyber-attacks. However, they are essential for the protection of our data. Cybersecurity in this matter is directed through biometric and facial identification methods to maintain the security of its users and their data (Ismagilova et al. 2020).

Given the facts, it is obvious that the new post-pandemic "normal" of COVID-19 will be associated with digital transformation and social changes. In this context of new post-pandemic world reality, it is also directly associated with new forms of work and the need for investment in remote access since this model of work reduces operational costs. Or even pondering a population increasingly adapted and immersed in a digital environment, this new "normal" has led many organizations to migrate their applications to the cloud, in addition to rethinking their cybersecurity policies, in order to avoid future threats (Costa, Peixoto 2020; De Las Heras et al. 2020).

Also, evaluating that home office systems will hardly return to physical offices, in this sense the management of cybersecurity in cloud computing since from the storage of data and information in this type of architecture, it is possible to access the web system from anywhere. In this scenario, which tends to be expanded and become more complex, managing cybersecurity in the cloud, reducing the risks of decentralizing data storage, and changing the perimeter that needs protection is becoming essential. This digital cybersecurity must consist of protection of several layers of security, where only the person responsible for the device can access the information through a combination of fingerprint, face recognition, PIN, lock pattern, and voice phrase, all of which are encrypted with AES-SHA 256 (Bhardwaj et al. 2016b, 2016c; Fields 2018; Gupta et al. 2020).

At the same time that Machine Learning is a fundamental technique of cybersecurity strategy, strengthening systems based on technology in the detection of malware, in addition to taking advantage of this innovation to generate alarms that allow social engineering attacks. In the face of these threats, in addition to preventing the occurrence of cyber-attacks, through layered security, seeking to identify and even contain complex attacks quickly (Bhardwaj et al. 2016b, 2016c; França et al. 2020a, 2020b).

Still reflecting on cybersecurity in remote work is seen as a major vulnerability, given that network security in an environment outside the organization, in general, is limited. Or even relating concerns about the security of remote work due to careless employees, unauthorized applications and mobile device security, or even including web application vulnerabilities, software/programming bugs, and inappropriate password practices. Reflecting on this aspect, a fundamental element of information cybersecurity is to control access to data, assuming that an employee should be allowed only the minimum level of access to data necessary to carry out their work, avoiding violations caused by data theft, malicious and accidental loss of data through phishing emails which are becoming more difficult to detect, or even through ATOs (Account Takeover) resulting in unauthorized transactions and exposure of confidential information. In contrast to these risks, AI software based on email cybersecurity can substantially mitigate the threat control that begins with phishing emails. As well as two-factor authentication (2FA) adding another layer of security to online accounts, evaluating that the technique is a feature that adds yet another "factor" to the normal login procedure, in order to verify the identity using two between three possible identifiers that the user knows (password, PIN number, zip code, among others); and something that this user is (facial recognition, fingerprints, retinal scan among others) (Braun et al. 2018).

Blockchain has a decentralized and encrypted organization, being more secure than other systems, which offers immeasurable advantages for data security, such as encryption of emails and messages in applications. Each block has the data + unique identification (called a hash) + hash information of the previous block to which it is connected. Pointing out that this hash changes and with the new identification, the other connected blocks had to be validated again (Fernández-Caramés and Fraga-Lamas 2018; Bernabe et al. 2019; Salah et al. 2019).

We are still considering that the combination of consensus algorithms and information replication allows guaranteeing the integrity and immutability of the data over time. In this sense, the technology guarantees a shared truth on which to work, avoiding the existence of a single point of failure, consisting of a solution to the challenges of cybersecurity since the information in the blockchain technology is all encrypted and who is capable of whether or not to validate a transaction are the computers themselves (via internal logic and algorithms) and not the people who operate these machines (Conoscenti et al. 2016; Khan and Salah 2018; Saberi et al. 2019).

7.6 CONCLUSIONS

A smart city makes use of technology as a way to help its residents to live happily through flood prevention, traffic reduction, garbage collection automation, among others. However, evaluating the smart devices and systems that allow smart city technologies to connect to the internet has vulnerabilities that can lead to concrete problems.

Since the vulnerabilities of IoTs and sensors are often disregarded, allowing unauthorized third parties to, for example, spy on a home using baby monitors or change the temperature of someone's connected thermostat. However, this becomes even more critical when considering that these devices and sensors have cybersecurity vulnerabilities that can control the water level in dams or alert residents of a flood episode or even radiation levels in locations close to a nuclear plant.

Still pondering that often, the issue of vulnerabilities, in general, is that it is not complicated, given that it is possible to be able to correct them by means of basic measures. However, it is clear that the threats caused by these vulnerabilities are real, and it is necessary to take the necessary corrective measures. Or relating that the challenge of maintaining cybersecurity in the cloud is a multifaceted process, which requires cooperation between the smart city (customer) and the cloud provider

It is also worth noting that AI is a strong ally of information security, especially due to its high availability and the possibility of learning all aspects, users, devices, and IoT and its aspects, capable of continuously discovering threats, making each increasingly difficult for certain cybersecurity vulnerabilities to progress.

REFERENCES

Alibasic, A., Al Junaibi, R., Aung, Z., Woon, W. L., & Omar, M. A. (2016). Cybersecurity for Smart Cities: A Brief Review. In *International Workshop on Data Analytics for Renewable Energy Integration* (pp. 22–30). Springer, Cham.

Alshehri, M., Bharadwaj, A., Kumar, M., Mishra, S., & Gyani, J. (2021). Cloud and IoT Based Smart Architecture for Desalination Water Treatment. *Environmental Research*, 195, 110812.

Al-Turjman, F. (2020). *Smart Cities Performability, Cognition, & Security.* Springer International Publishing, Cham.

Arya, K. V., Bhadoria, R. S., & Chaudhari, N. S. (Eds.). (2018). *Emerging Wireless Communication and Network Technologies: Principle Paradigm and Performance.* Springer, Berlin.

Beebe, N. H. (2019). *A Complete Bibliography of Publications in Science of Computer Programming.* Elsevier, Amsterdam.

Bernabe, J. B., Canovas, J. L., Hernandez-Ramos, J. L., Moreno, R. T., & Skarmeta, A. (2019). Privacy-Preserving Solutions for Blockchain: Review and Challenges. *IEEE Access*, 7, 164908–164940.

Bhardwaj, A., Al-Turjman, F., Kumar, M., Stephan, T., & Mostarda, L. (2020). Capturing-the-Invisible (CTI): Behavior-Based Attacks Recognition in IoT-Oriented Industrial Control Systems. *IEEE Access*, 8, 104956–104966.

Bhardwaj, A., & Goundar, S. (2018). Reducing the Threat Surface to Minimise the Impact of Cyber-Attacks. *Network Security*, 2018(4), 15–19.

Bhardwaj, A., & Goundar, S. (2019). A Framework for Effective Threat Hunting. *Network Security*, 2019(6), 15–19.

Bhardwaj, A., Subrahmanyam, G. V. B., Avasthi, V., & Sastry, H. (2016a). Security Algorithms for Cloud Computing. *Procedia Computer Science*, 85, 535–542.

Bhardwaj, A., Subrahmanyam, G. V. B., Avasthi, V., & Sastry, H. G. (2016b, January). *Solutions for DDoS Attacks on Cloud*. In *2016 6th International Conference-Cloud System and Big Data Engineering (Confluence)* (pp. 163–167). IEEE, New York.

Bhardwaj, A., Subrahmanyam, G. V. B., Avasthi, V., Sastry, H., & Goundar, S. (2016c, October). DDoS Attacks, New DDoS Taxonomy and Mitigation Solutions—A Survey. In *2016 International Conference on Signal Processing, Communication, Power and Embedded System (SCOPES)* (pp. 793–798). IEEE, New York.

Braun, T., Fung, B. C., Iqbal, F., & Shah, B. (2018). Security and Privacy Challenges in Smart Cities. *Sustainable Cities and Society*, 39, 499–507.

Cavada, M., Hunt, D., & Rogers, C. (2017). *The Little Book of Smart Cities*. The University of Birmingham, Birmingham.

Choudrie, J., Kurnia, S., & Tsatsou, P. (Eds.). (2017). *Social Inclusion and Usability of ICT-Enabled Services*. Routledge, Abingdon.

Colaković, A., & Hadžialić, M. (2018). Internet of Things (IoT): A Review of Enabling Technologies, Challenges, and Open Research Issues. *Computer Networks*, 144, 17–39.

Coletta, C., Evans, L., Heaphy, L., & Kitchin, R. (Eds.). (2018). *Creating Smart Cities*. Routledge, Abingdon.

Conoscenti, M., Vetro, A., & De Martin, J. C. (2016). Blockchain for the Internet of Things: A systematic literature review. In *2016 IEEE/ACS 13th International Conference of Computer Systems and Applications (AICCSA)* (pp. 1–6). IEEE, New York.

Conti, M., Dargahi, T., & Dehghantanha, A. (2018). Cyber Threat Intelligence: Challenges and Opportunities. *Cyber Threat Intelligence* (pp. 1–6). Springer, Cham.

Costa, D. G., & Peixoto, J. P. J. (2020). COVID-19 Pandemic: A Review of Smart Cities Initiatives to Face New Outbreaks. *IET Smart Cities*, 2(2), 64–73.

De Las Heras, A., Luque-Sendra, A., & Zamora-Polo, F. (2020). Machine Learning Technologies for Sustainability in Smart Cities in the Post-COVID Era. *Sustainability*, 12(22), 9320.

Farahat, I. S., Tolba, A. S., Elhoseny, M., & Eladrosy, W. (2019). Data Security and Challenges in Smart Cities. *Security in Smart Cities: Models, Applications, and Challenges* (pp. 117–142). Springer, Cham.

Farsi, M., Daneshkhah, A., Hosseinian-Far, A., & Jahankhani, H. (Eds.). (2020). *Digital Twin Technologies and Smart Cities*. Springer International Publishing, Cham.

Fernández-Caramés, T. M., & Fraga-Lamas, P. (2018). A Review on the Use of Blockchain for the Internet of Things. *IEEE Access*, 6, 32979–33001.

Fields, Z. (Ed.). (2018). *Handbook of Research on Information and Cyber Security in the Fourth Industrial Revolution*. IGI Global.

França, R. P., Iano, Y., Borges, A. C., Monteiro, R. A., & Estrela, V. V. (2020a). A Proposal Based on Discrete Events for Improvement of the Transmission Channels in Cloud Environments and Big Data. *Big Data, IoT, and Machine Learning: Tools and Applications*, 185.

França, R. P., Iano, Y., Monteiro, A. C. B., & Arthur, R. (2020b). Better Transmission of Information Focused on Green Computing Through Data Transmission Channels in Cloud Environments with Rayleigh Fading. *Green Computing in Smart Cities: Simulation and Techniques* (pp. 71–93). Springer, Cham.

França, R. P., Iano, Y., Monteiro, A. C. B., & Arthur, R. (2020c). A Proposal of Improvement for Transmission Channels in Cloud Environments Using the CBEDE Methodology. *Modern Principles, Practices, and Algorithms for Cloud Security* (pp. 184–202). IGI Global.

França, R. P., Iano, Y., Monteiro, A. C. B., & Arthur, R. (2020d). Improvement of the Transmission of Information for ICT Techniques Through CBEDE Methodology.

Utilizing Educational Data Mining Techniques for Improved Learning: Emerging Research and Opportunities (pp. 13–34). IGI Global.

França, R. P., Iano, Y., Monteiro, A. C. B., & Arthur, R. (2020e). A Review on the Technological and Literary Background of Multimedia Compression. *Handbook of Research on Multimedia Cyber Security* (pp. 1–20). IGI Global.

França, R. P., Iano, Y., Monteiro, A. C. B., & Arthur, R. (2020f). Intelligent Applications of WSN in the World: A Technological and Literary Background. *Handbook of Wireless Sensor Networks: Issues and Challenges in Current Scenario's* (pp. 13–34). Springer, Cham.

França, R. P., Iano, Y., Monteiro, A. C. B., & Arthur, R. (2020g). Big Data and Cloud Computing: A Technological and Literary Background. *Advanced Deep Learning Applications in Big Data Analytics* (pp. 29–50). IGI Global.

França, R. P., Monteiro, A. C. B., Arthur, R., & Iano, Y. (2020h). An Overview of Deep Learning in Big Data, Image, and Signal Processing in the Modern Digital Age. *Trends in Deep Learning Methodologies: Algorithms, Applications, and Systems*, 4, 63.

França, R. P., Monteiro, A. C. B., Arthur, R., & Iano, Y. (2020i). An Overview of Internet of Things Technology Applied on Precision Agriculture Concept. Precision Agriculture Technologies for Food Security and Sustainability. *Precision Agriculture Technologies for Food Security and Sustainability*, 47–70. IGI Global, Hershey.

França, R. P., Monteiro, A. C. B., Arthur, R., & Iano, Y. (2020j). An Overview of Blockchain and its Applications in the Modern Digital Age. *Security and Trust Issues in Internet of Things: Blockchain to the Rescue*, 185. CRC Press.

França, R. P., Monteiro, A. C. B., Arthur, R., & Iano, Y. (2020k) An Overview of the Integration between Cloud Computing and Internet of Things (IoT) Technologies. *Recent Advances in Security, Privacy, and Trust for Internet of Things (IoT) and Cyber-Physical Systems (CPS)* (pp. 1–22).

Goundar, S., & Bhardwaj, A. (2018). Efficient Fault Tolerance on Cloud Environments. *International Journal of Cloud Applications and Computing (IJCAC)*, 8(3), 20–31.

Gupta, B. B. (Ed.). (2018). *Computer and Cybersecurity: Principles, Algorithm, Applications, and Perspectives*. CRC Press.

Gupta, B. B., Perez, G. M., Agrawal, D. P., & Gupta, D. (2020). *Handbook of Computer Networks and Cyber Security*. Springer Science and Business Media LLC, New York.

Iqbal, S., Kiah, M. L. M., Dhaghighi, B., Hussain, M., Khan, S., Khan, M. K., & Choo, K. K. R. (2016). On Cloud Security Attacks: A Taxonomy and Intrusion Detection and Prevention as a Service. *Journal of Network and Computer Applications*, 74, 98–120.

Ismagilova, E., Hughes, L., Rana, N. P., & Dwivedi, Y. K. (2020). Security, Privacy and Risks Within Smart Cities: Literature Review and Development of a Smart City Interaction Framework. *Information Systems Frontiers*, 1–22.

Jayapandian, N. (2019). Threats and Security Issues in Smart City Devices. *Secure Cyber-Physical Systems for Smart Cities* (pp. 220–250). IGI Global.

Kamyab, H., Klemeš, J. J., Van Fan, Y., & Lee, C. T. (2020). *Transition to Sustainable Energy System for Smart Cities and Industries*. Elsevier, Amsterdam.

Khan, J. Y. (2019). Introduction to IoT Systems. *Internet of Things (IoT): Systems and Applications*. CRC Press, London.

Khan, M. A., & Salah, K. (2018). IoT Security: Review, Blockchain Solutions, and Open Challenges. *Future Generation Computer Systems*, 82, 395–411.

Kim, D., & Solomon, M. G. (2016). *Fundamentals of Information Systems Security with Cybersecurity Cloud Labs: Print Bundle*. Jones & Bartlett Learning.

Komninos, N. (2019). *Smart Cities and Connected Intelligence: Platforms Ecosystems and Network Effects*. Routledge, Abingdon.

Kunz, M., Puchta, A., Groll, S., Fuchs, L., & Pernul, G. (2019). *Attribute Quality Management for Dynamic Identity and Access Manage*.

AI and Machine Learning include natural language processing technologies, neural networks, and Deep Learning. The trend is for these technologies to evolve with the aggregation of more advanced systems that are capable of understanding, learning, predicting, adapting, and potentially even operating autonomously (Flasiński 2016).

Machine Learning and AI are the most viable and effective ways of evolution to identify and respond to automated and efficient incidents. The algorithms are used to collect all available data to create the best possible informational scenario for identifying patterns and developing ideas. Specifically, in information security, its greatest use is to remove from a large amount of information that is processed today useful and actionable information (Rebala et al. 2019).

AI and Machine Learning can be useful to make the environment safer. These technologies help IT, and security professionals more quickly identify risks and anticipate problems before they occur. When used in favor of protecting a given set of information or a system, the Machine Learning technique guarantees continuous improvement of the protective barriers and networks that make up the system's connections. Machine Learning allows protection technology to take advantage of potential threats, as it has already managed to identify weaknesses in the system on its own and correct them before an external mechanism tries to carry out this process to force a breach of security. Besides, when the algorithm identifies a threat, it can quickly act to prevent data loss. AI software can also easily adapt to the constantly evolving threat landscape based on Big Data insights. In the financial sector, for example, companies can use AI and Machine Learning techniques to track transactions in real-time and establish predictive models that can show the likelihood of a transaction being fraudulent (Fraley and Cannady 2017; Handa et al. 2019).

This trend needs to assess a large number of scenarios in which AI and Machine Learning can generate value for the business and experience the greatest impact. At the same time, organizations need to prepare a security strategy that adapts to this new scenario, which may include collecting data from different devices to create increasingly intelligent systems, such as multiple security platforms and smart object sensors. In addition to identifying threats, this technology also allows automating actions, which, in short, means that it is possible to stop attacks before they start (Hutter et al. 2019).

Therefore, this chapter has the objective of providing an updated overview of Machine Learning, addressing fundamental concepts its relationship with Cyber Security facing the Internet of Things (IoT), with a concise bibliographic background, categorizing and synthesizing the potential of technology.

8.2   IoT CONCEPT

The IoT can be included in the list of digital security trends, given that the technological concept relates that all devices are connected. For this reason, an action performed on one device can be identified on another, so with intelligent data processing technologies, it is possible to identify suspicious actions. The IoT concept relates that any device, even those imaginable, can have some way of connecting to the internet, from a refrigerator, electronic gates, thermostats, microwaves, televisions, phones, security cameras to a smartphone, among others (França et al. 2020a–2020f).

By allowing all objects to communicate with each other, IoT makes the digital world and the physical world connect and influence each other. Representing an evolution in the role of IT, since the entire infrastructure is used as a basis for connection, storage of information generated between devices, and also communication. IoT refers to a digital information ecosystem in which several devices are intelligently interconnected in productive connectivity. Dealing with a new stage of the digital revolution, in which the internet is able to support elements far beyond conventional ones (França et al. 2020a–2020f).

The current technological reality allows and encourages the creation of a network of interconnected physical objects (IoT) by a highly intelligent system, in which one device dialogues with another in a large systemic and productive chain. In this information ecosystem, which is based on the internet, it provides a universe of practical and applicable possibilities, both from the users' point of view and in the world of business entrepreneurs (Al-Turjman 2019).

However, the technology still faces some fears, especially when it comes to security aspects, with regard to the structure being harmoniously interconnected. It is an extremely positive idea and in line with future trends. But it is enough for a cyber-criminal to invade this system to have access, at once, to all the data and the functioning of a large and complex organization, for example (França et al. 2020a–2020f).

Intelligent things are physical objects that go beyond the execution of rigid programming models to explore aspects of AI as a way to offer advanced behaviors and interact more naturally in different environments and with users. This innovation promotes profound advances for autonomous vehicles, drones, and even robots, offering improved capacity for many other objects, which allow these things to connect to the internet and increase interaction with users (Al-Turjman 2019).

IoT has become an indispensable ingredient for organizations to improve productivity, product quality, services, and even business strategies. However, the fear of full use of technology lies in the fact that IoT devices can be invaded by forming "enslaved" networks of compromised elements. Or, even if the organization does not have control over all IoT devices on the network, it means that there may be security vulnerabilities of which there is no awareness. In other words, these devices, uncontrolled, represent a huge opportunity for cyber-intrusion and a great digital risk for the company. To minimize this risk, it is necessary to identify in a controlled and systematic manner all assets connected to the network, adapting and framing the cybersecurity strategy (Strong 2016).

The adoption of IoT is derived from the possibility of analyzing data as a basis for companies to be able to improve their business strategies and their operation as a whole. Representing that IoT is an essential technology to provide valuable information that can be used in the decision making process, development of new products, services, still allowing the automation of several processes and, thus, contributing to the reduction of costs and greater operational efficiency, and even improvement of day-to-day activities (França et al. 2020a–2020f).

At the same time, Cloud technology is essential to achieve adequate storage and processing power for IoT applications to operate. Given the intense data traffic, security is paramount, which is not recommended to put this volume of information in risky situations, such as virus attacks or malicious people. Therefore, Cloud Computing services provide guarantees of digital security related to continuity, backup and disaster recovery capacity, and even digital privacy when encrypting data. The IoT era is associated with Cloud Computing services since the Cloud facilitates the entire IoT process with automated tools, working continuously, in addition to being a fast, available, and flexible service (França et al. 2020a–2020f, 2021).

IoT perfectly combines hardware and software capable of delivering billions of connected objects, which makes life easier for users in several aspects, in addition to creating possibilities that impact business models and the performance of organizations. The great advantage of the IoT is in the connectivity of mobile devices with any object with the internet. However, it has little to deliver if it is not integrated with other technologies, given the example with Cloud Computing as a key factor facilitating the storage of all the data that these objects generate (Shovic 2016).

AI is another technology that, integrated with IoT, is able to deliver even more possibilities, derived from statistical analysis tools and the metrics used, transform digital perception, and generate improvements. Allowing learning to connected IoT devices, making the work environment optimized with data debugging, and providing more efficient use of both devices and data generated, collected, stored, and processed. In an applicable practical context, there is an autonomous vehicle that, equipped with AI, perceives the environment through IoT sensors, without the need for a human driver. Thus, innumerable possibilities are allowed from the convergence between IoT, Cloud Computing, and AI (França et al. 2020a–2020f).

Considering a world increasingly rich in integrated platforms, the traditional metrics and statistical analysis tools will not be enough, considering the need for systems capable of developing some type of knowledge and that have digital perception. Also relating the ability to learn machines in natural processes of the coexistence relationship between humans and machines. Given that AI purifies the data and allows its use in a more efficient and even profitable way, exemplifying the use of the intelligent agent, which perceives the environment through IoT sensors and acts through actuators (Jackson 2019).

Thus, the union between IoT and AI also allows the development of new business models, well-structured considering more complex analyzes, as a result of the interactions between the various IoT devices and connected equipment. As well as the intersection between Cloud Computing, AI, and IoT allows innovations in support of strategic management. As IoT objects learn and integrate with the actions of users, they become agents of change. This technological confluence provides a digital opportunity to mature the use of technological trends for the evolution of human society (Al-Turjman 2019).

8.2.1 IoT Aspects Security

The reduction in the size of electronic components and the decrease in their prices have gradually allowed the addition of intelligent resources on machines and connect them to the internet, i.e., IoT, from those used to cool the air in homes or the workplace, to those that allow the operation of trains, or those applications in the health area, among others (Handa et al. 2019).

Although IoT has a direct impact on the efficiency of organizations and society in general, it also brings new risks since many problems arise since the devices do not have integrated security. IoT technology is relatively still new, which means that security is not fully mature. That is, in large part, representing that the adoption of these solutions is not only gaining benefits but also a greater number of points vulnerable to cyber-attacks (Mylrea et al. 2017; Roopak et al. 2019).

From a historical point of view, major DDoS cyber-attacks (Distributed Denial of Service) arose out of an IoT vulnerability, either by taking down a company that makes servers available, or botnet of hijacked IoT devices, or taking down the world's largest sites. Emphasizing that the problem is worse than that, as vulnerable IoT devices can be hacked and exploited in giant botnets that threaten even properly protected networks (Roopak et al. 2019).

Considering that for a botnet composed mainly of IoT devices, generating a representative amount of traffic takes some time to be analyzed and reveal which mitigation tools will be the most efficient to avoid it. Resulting in the objective that among so many devices that the attacker had access to, it is not even possible in time to employ common tactics that amplify the negative impact (Babu et al. 2019).

Or even reflecting on the vulnerability of malware that can be installed on computers, looking for other vulnerable devices spreading the malware further. IoT has low latency and low km^2 connection capacity, which represents networked devices and an unprecedented amount of data circulating. Therefore, invasion opportunities are due to vulnerabilities within the IoT environment. This scenario is even worse when considering that an unprotected device (node) is enough for attackers to gain access to the network, and from then on, even the devices with some protection end up being threatened. And even more deeply, the data that transit in this environment has great chances of being captured (Tweneboah-Koduah et al. 2017).

This is a result of what was briefly highlighted, which is that solutions for security control and threat identification do not yet exist (or at least the most sophisticated solutions are scarce) for IoT devices. Since if a device is identified, it reflects that the network recognizes its identity. And so, attackers can still try to infiltrate another device on the network, since in IoT networks, there is no guarantee that each device is unique and legitimate (which would be a way to protect the network). It also represents a problem of digital visibility for the security team, who cannot easily know all the devices that belong to the network. On the other hand, the device, whatever it is, still serves as a gateway to access internal networks (Tweneboah-Koduah et al. 2017; Babu et al. 2019).

Still evaluating the modern factor known as social hacking as one of the most efficient ways to achieve illegal access. However, the solution is to raise awareness, especially in companies, considering a challenge for all employees, affecting a crucial point, which is passwords. Since DDoS attacks, passwords, and standard users are dangerous, which requires good practices and mandatory password and user exchange for any IoT device employed. Given that the problems with IoT security are related to the detection of intrusions, a factor that the technology cannot identify when they suffer attacks on their IoT devices (Liu et al. 2019).

The technology lacks the adoption of specific and robust guidelines to address the lack of secure update mechanisms, including the lack of firmware validation on the IoT device, lack of secure delivery (not encrypted for data in transit), lack of anti-rollback mechanisms, and even lack of notifications of security changes. However, the cybersecurity factor is not always an IoT technology issue, since in some cases, the physical location of IoT devices makes updating, repairing, and even replacing a significant challenge (Thames & Schaefer 2017).

Private networks owned only for IoT devices can serve as a solution, considering that the interaction with the main network is even more restricted, which means that the most important files are isolated. This is a way to improve not only IoT security but also the entire network, considering that each of the private networks can be configured and managed in an isolated and centralized way. Resulting in more control of everything that happens in each one of them (Thames & Schaefer 2017).

IoT devices can be divided into two groups according to their functions, i.e., those that collect information through environmental sensors to transmit information constantly, or those devices developed to receive instructions via the internet and perform some activity in the place where they are located, as well as devices that can perform both functions. In both cases, two points deserve concern, i.e., privacy, given who can access the data collected by any device and for what, and security, who can command the actions of the device (Tran & Misra 2019).

The lack of encryption or access control of confidential data anywhere in the IoT ecosystem, including data transfer and storage at rest, in transit, or during processing. It also deserves attention to secure storage, ensuring that data remains secure during transfer. In this sense, one of the resources used to eliminate vulnerabilities may be Blockchain. Or the lack of security support in IoT devices deployed in production, including asset management, update management, secure decommissioning, systems monitoring, and response capabilities. Considering that IoT devices can be small, inexpensive, and deployed in large numbers, but that does not mean that there is no need to manage them, which is the opposite. This makes management even more important. Therefore, if there is a vulnerability, which cannot be seen and identified, there is a great associated risk (Khan & Salah 2018).

Some reasons why vulnerabilities exist in IoT devices are more frequent, such as the lack of compliance and security concerns on the part of the manufacturers, slowness or lack of updates on the devices, and even the ignorance of the vulnerabilities on the part of the user. These threats result in several types of attacks, such as health data leakage, sensitive data hijacking, and even industrial espionage. Still discussing the fact of what makes IoT devices a preferred target for cybercriminals is the fact that it consists of a complex ecosystem of information circulating, as it is collected and transmitted all the time in different places and types of users. Among the current data are many personal and sensitive elements, such as financial information. It is the degree of sensitivity of these assets (address, social registers, and even health data of users) that cybercriminals are interested in, correlating this risk with digital security involving IoT technology with regard to data privacy (Thames & Schaefer 2017; Meneghello et al. 2019).

Which means that any IoT device connected to the network needs to be identified and evaluated, i.e., efficient asset control. If there is a vulnerability that cannot be eliminated, the IoT device must be replaced, or as new vulnerabilities appear, it is necessary to keep this updated to combat them. If the software is not prepared for this, it is always exposed to new cyber-attacks. Reflecting that generally, the main security challenges in IoT result from the systems and devices themselves, given their limited or even nonexistent defense resources, not having the security functions, at least traditional that would limit invasions (Meneghello et al. 2019).

Or even considering that there is still no standard protocol for IoT communication and security, reflecting the risk of malware infections, or even citing endpoints that represent the connection of devices to the network and can serve as an entry for external threats. In other words, IoT security holes in the attack surface and intrusions become greater, given the direct action linked to DDoS attacks, which causes servers, systems, devices, and networks to be brought down by massive access. Acting from an infected master computer, which powers thousands of zombie machines that start accessing the network, resulting in real users being unable to access it, and the attackers start their invasion (Thames & Schaefer 2017).

That is why bank details (sensitive data while remaining unprotected) are stolen from an intelligent refrigerator from which a purchase is made. Or, because of the location system of an autonomous delivery car, criminals map a particular route and plan thefts. To deal with IoT and its security factors, it is necessary to equip each and every IoT device with protection solutions (practices such as double authentication, added steps and measures that limit access, use of strong passwords, and even constant updating, among others) for digital tools, and later, strengthen the gateways that connect these devices. Finally, awareness of the vulnerabilities that can be exploited is the major challenge around IoT today, as well as a better understanding of this scenario, which helps to develop measures that increase information security (Miloslavskaya & Tolstoy 2019).

After all, while the IoT has a huge applicable earning potential, it also opens the door to new vulnerabilities, which must be the constant balance that technology must achieve. Aligned with the IoT becoming more and more a standard, it is necessary to be very concerned with your digital security (Miloslavskaya & Tolstoy 2019).

8.3   MACHINE LEARNING CONCEPT

The large volume of data and information available on computers and the network has changed the direction of AI, a concept in which machines are "trained" so that they can learn to rationalize processes as human intelligence performs, observing processes and learning from them and their respective changes, influencing the ability of malicious software to invade systems and the resulting digital protection responses. It is in this context that Machine Learning emerges as one of the most promising ways of applying AI for the development of information security systems. Technology is a set of equations, statistical models, and algorithms based on AI that, when applied specifically in information security, its greatest use is to remove, from a large amount of processed information, useful and actionable information (Cielen et al. 2016).

Machine Learning is defined as the ability of computers to learn to perform tasks without having been explicitly programmed to do so. It is a way of creating

AI from the collection and analysis of data. That is, using mathematical techniques on large data sets, the algorithms build behavioral models and use them as a basis to perform actions and make predictions based on new input data, the results of which allow the machine to "learn" a certain task to perform it by itself or manage to determine a future situation or information based on observed patterns (Monteiro et al. 2020).

Machine Learning involves creating algorithms that analyze data, learn from that data, and then apply what they learn to make decisions. In a practical, applicable context, technology is employed in a music streaming service. In order for the digital service to make a decision about which new songs or artists to recommend to the user, the learning algorithms associate the user's preferences with others who listen to the same artists or music genres (Raschka & Mirajalili 2019).

Machine Learning is the data analysis method that automates the development of analytical models, allowing computers to find more strategic insights for decision making. In other words, it is related to Machine Learning and, from there, having the best instructions. Being programmed from intelligence that analyzes the data and finds the best ways to resolve issues, that is, it is possible to develop models capable of analyzing larger and more complex data and providing faster and more accurate results that can lead to better decisions and smart actions all the time (Simeone 2017).

Machine Learning applies to all types of automated tasks and extends to various sectors of society. Corresponding to "learning," it means progressively better in the execution of functions over time. Technology is on the rise because it has the ability to take solutions and find relevance where a human being would not see, i.e., generate insights and intelligent decisions (Rebala et al. 2019).

Machine Learning applied to IoT refers to a technology in which, through predictive algorithms, it becomes possible for systems to be able to learn from the patterns generated by the inputs (IoT devices) of information and be able to adapt to changes that occur along the time, without having to change the parameters created initially. Since the more the processes are carried out, the more refined the process becomes, considering that this will improve the practice according to what it executes. Thus, human interference is minimal in the process, being present in the routines of people and society, but it does it in such a silent way that it becomes invisible to the eyes of ordinary users (Stamp 2017).

Machine Learning has characteristics related to AI in the use of large databases for the execution of its activities; and the need for resources that allow the analysis of this large database. In practical terms, the devices are capable of analyzing documents, analyzing and improving the electrical system to save energy costs, improve customer service, recommend the right product for each consumer profile, among other possibilities (França et al. 2020a–2020f).

Through Machine Learning, it is possible to optimize results and perform an intelligent crossing of a universe of internal and external data involving an organization. That is, it has never been possible to know in such a detailed way the entire context of a business in the market, with information capable, generate value for her activities. Or it can automate processes that are highly repetitive or that require a lot of human effort and can be automatically replaced by machines, software, and AI robots (Sterne 2017; Ullah et al. 2020).

Through algorithms, they are able to significantly increase the speed of the service and, also, improve the results to levels of excellence and precision. These intelligent technological tools are capable of providing much more qualified diagnoses than doctors, legal analyzes more improved than lawyers, more accurate calculations than accountants, management plans more modern than administrators, among others. In this case, it is not a question of purely mechanized functions, i.e., replacing competencies previously exclusive to the human intellect that imply the end of the need for these professionals. On the contrary, they will serve as an object to improve the results in their respective areas, in a harmonic interaction between people and machines (Zhou et al. 2017; Alpaydin 2020).

The main feature that differentiates Machine Learning devices from traditional automation machines is their ability to perform intelligent actions without the need for having been previously programmed, i.e., unlimited learning. This learning can be supervised when there is greater human interaction in the process, or unsupervised, when the machine achieves a total level of autonomy in its performance, regardless of the complexity of its task. This independence in performance is through cognitive computing systems, having the ability to proactively absorb information from the environment in which they are installed. Still considering the ability to analyze these data, process them, and make assertive decisions for the evolution of the work (Simeone 2017; Zhou et al. 2017; Alpaydin 2020).

Machine Learning solutions are able to increase profitability not only by reducing costs but by considerably increasing productivity, in several ways, still correlating that intelligent and automatic data analysis also generates savings. Still reflecting that in addition to making processes more agile, there is more effectiveness in actions. The importance of Machine Learning comes from its main advantages related to the ability to analyze more complex data, agility in performing tasks, building more accurate models, reducing the need for a human component in repetitive tasks, among others (Simeone 2017; Rebala et al. 2019).

8.3.1 Types of Learning

Machine Learning programming is subdivided into Supervised Learning, establishing models of known data entry and forecast output. The computer is "taught" what each entry is (which is the label), and this learns what characteristics of those entries make them the that are (Monteiro et al. 2020).

Supervised Learning determines a learning algorithm from a set of known data to classify information. In addition, in parallel to this categorization system, the system can still recall previous data entries to make predictions and deductions based on the groups of information already incorporated (Monteiro et al. 2020).

According to the type of the result of the algorithm, it is possible to classify it between classification algorithms, exemplifying an entry between two possible types, that is, dog or non-dog; teaching the machine to recognize what a bicycle is, between patterns and similarities. Color and size may vary, but the machine learns that a bicycle has pedals, two wheels, handlebars, and other key elements. Or even a regression algorithm, occurring when the result is numeric. An applicable practical context of this technology is an algorithm that calculates the value of a home based on characteristics such as the number of rooms, built area, location, and year of construction. And from examples of similar houses, the algorithm learns to price new houses (Simeone 2017).

Machine Learning programming is also subdivided into *Unsupervised Learning*, identifying patterns and structures hidden in data entry, referring to programming that finds hidden patterns or special structures in the data. Also called "Clustering," this specification allows building estimates for complex information and without records in the system since the crossing of the most successful responses sets the condition of "knowledge" of the machine (Neapolitan & Jiang 2018; Monteiro et al. 2020).

Unsupervised Learning, the algorithm is taught to separate data into similar groups without saying what these groups are. Similarly, with image classification, there is a model that has learned to classify images between two different groups. However, when receiving a new image, based on its attributes, it identifies which group it belongs to, without necessarily knowing what that group is Neapolitan and Jiang (2018).

This technique is a type less used by companies because the machine begins to analyze data alone and to identify patterns, learning to separate what is a can from a bottle, for example. However, as it is Machine Learning by itself concepts that it has never seen before, the process takes longer. An applicable practical scenario is the "matches" in relationship apps and suggestions for connections in a professional social network focused on generating connections and professional relationships (Hsu et al. 2018).

Reinforcement learning is a way of teaching the machine which action to prioritize given a given situation, making it possible to link rewards and punishments to possible results and, considering them in the right way, teach the priority level of each goal. This way, it is taught which actions the computer should prioritize. This type of reinforced learning is similar to what happens with humans at a young age (children). It is related to teaching based on experience, in which the machine must deal with what went wrong before and look for the right approach. An applicable practical scenario is a recommendation on a video sharing platform, given that after the user watches a video, the platform will show similar titles that he believes the user will also like. However, if that user watches the recommended and does not finish it, the machine understands that the recommendation was not a good one and will try another approach next time (François-Lavet et al. 2018).

The similarity present in the three types of learning is related to Machine Learning to infer something based on their past experiences. Considering the difference between unsupervised and Supervised Learning is that in the first approach, learning occurs with unlabeled data. That is, do not is taught to the computer what that entry is Monteiro et al. (2020) and França et al. (2020a–2020f).

8.3.2 DEEP LEARNING

Deep Learning is just a subset of Machine Learning, which technically works similarly. However, Machine Learning models become progressively better in whatever role but still need some guidance. If an algorithm returns an inaccurate forecast, adjustments will need to be made. However, with a Deep Learning model, algorithms can determine for themselves whether a forecast is accurate or not (Goodfellow et al. 2016).

A Deep Learning model is designed to continuously analyze data with a logical structure similar to how a human would conclude. For this, it uses a layered structure of algorithms called artificial neural network (RNA), inspired by the biological neural network of the human brain. This makes the machine's intelligence much more efficient than that of standard Machine Learning models. In a simplified way, Machine Learning and Deep Learning are two subsets of AI that have revolutionized the relationship between man and technology (Goodfellow et al. 2016; Charniak 2019; Monteiro et al. 2020).

8.4 DISCUSS

Machine Learning can be considered the revolution for information security since machines are trained to recognize patterns and can use this feature to identify the commands used by cybercriminals, preventing illegal access to corporate data. It is also possible to identify the algorithms used by viruses and malware to steal data, predict cyber-attacks, and reveal whether malicious programs can invade a system using just a few lines of code.

Relating that the nonlinear models of security classification, in general, come from all data that is received, treated, and processed so that the statistical models have different parameters on which is possible base predictions. In cases of use of Machine Learning algorithms, the use of algorithm stacks is recommended. As a combination of models using random classification forest (establishing the probability of something being malicious based on factors such as a domain, country, origin, and among other characteristics), and also using clustering models (grouping this information into types of threat such as phishing, botnet, ransomware, among others).

Whereas billions of devices (IoT) are currently in use, and judging that, a considerable percentage are vulnerable, connected to an extremely hostile environment. Still evaluating that it takes a few minutes for a cybercriminal to find thousands of vulnerable devices and compromise IoT devices that often become the gateway to other more serious breaches on networks.

The perceived need for AI technologies and strands (Machine Learning) is driven by the digital security complications introduced by IoT devices. This large number of devices connected to the internet means that the need to ensure a broader scope of digital security, given that these IoT devices, in general, have a low capacity to protect your IoT applications.

Also relating that Big Data technology served as a kind of learning database that allowed AI to learn autonomously and comprehensively, reaching a large amount of information for different data, patterns and relationships can be perceived through its processing analytical. Assessing that to deal with the security challenges of IoT devices, new technologies are needed to "discover and understand the threats that are active in the IT infrastructure." In addition, when the Machine Learning algorithm identifies a threat, it can quickly act to prevent data loss and can easily adapt to the constantly evolving threat landscape based on Big Data insights.

The most significant benefit of using AI and Machine Learning is to reduce the amount of time and effort required to investigate an alert. The benefits related to IoT technology derive from the reduction in the number of false-positive alerts, automation of tasks in the investigation, decision making, and remediation processes; searching and locating cyber-attacks before they do damage; and even improve coordination between networks, security, and operations. The most likely independent of the IoT environment employing AI and Machine Learning technology is to classify the processes most likely to contain attacks as a kind of digital quarantine. After that, the correction of cyber-attacks, the investigation of alerts, the classification of risks, the prioritization of alerts, and even the aggregation of forensic data.

Machine Learning is the most viable and effective evolution path for identifying and responding to automated and efficient cyber incidents, given that the algorithms employed collect all available data creating the best possible informational scenario for identifying patterns and developing countermeasures.

In network security, it is possible to use traffic profiles to recognize potential threats, in addition to conducting user behavior analysis to detect internal threats, in addition to using this technology to filter spam and identify malware. Machine Learning can also perform pattern recognition in the network flow, analyze historical data, logs, signatures, and other sources to identify trends. Therefore, the main advantage of using Machine Learning is its ability to process and analyze huge volumes of data quickly.

Systems with this technology are able to understand and even change future behaviors, leading to the creation of smarter programs and devices against cyber-attacks, driven by the combination of three characteristics related to greater data processing power, advanced intelligent algorithms, and the collection of large volumes of data. Still considering these characteristics combined with the combined possibility of using mathematical calculations and specific algorithms to guide and observe, leaving evidence the understanding of the entire digital infrastructure related to information security.

Still used by Cloud technology to consolidate intelligence and Machine Learning aimed at detecting anything suspicious. Considering that the Cloud approach aggregates in a Cloud data server all terminals that depend on IoT devices, in this sense, the input is analyzed (Machine Learning) to determine patterns and detect malicious behavior.

Machine Learning and automatic learning of components critical to IoT security are distributed around the IoT structure. Given the occurrence of a cyber-attack, it is reacted in real-time since most systems that rely on Machine Learning and digital behavior analysis will collect information about the network and connected devices (IoT) and then search for anything that is out of the ordinary. Developing a solution approach that potentiates an analyst's view of increasing the detection of cyber-attacks, reducing false alarms, and increasing accuracy.

Still pondering that the Machine Learning technique guarantees continuous improvement of the protective barriers and the networks that make up the system connections, present in a cybersecurity solution, which can go deeper into the history of security data recognizing the possibility of a specific attack based on your variables and relationships and predict, based on that knowledge, the next cyber-attack.

Taking advantage of the limited functionality of IoT devices since they are designed to perform a limited set of functions, thus, with sufficient Machine Learning and data, it becomes quite easy to identify anomalous behaviors.

Machine Learning allows protection technology to be at an advantage over potential cyber threats since it has managed to identify weaknesses in the system on its own and correct them before an external mechanism tries to carry out this process to force a breach of protection digital.

Through the collection of metadata from different endpoints and the definition of a range of behavior for each IoT device, Machine Learning can block arbitrary behaviors, as the detections get even better as more data is collected. This technique is an important step in IT and digital security since they remove from a passive role of containment to an active position in the dam of cyber-invasions. Cybersecurity attacks caused by the fragility of IoT network perimeters and the scarcity of security qualifications make cp, qi AI solutions act as a fundamental tool to eliminate these gaps and stealth threats in IoT cybersecurity infrastructures.

Machine Learning has the capacity to process millions of daily accesses, filtering data, and transmitting only potential threats, showing a higher attack detection rate and, at the same time, a reduction in false positives. Machine Learning, in this sense, appears as an effective solution to face cyber threats that are constantly changing.

This technology is essential for detecting and stopping attacks targeting IoT user devices and successfully protecting data and other high-value assets, providing a more efficient investigation and responding more quickly to sneak attacks that were not identified by the network's perimeter defense systems, and even identifying attacks that use IoT devices as a gateway as a security gap. However, it is worth noting that Machine Learning is only as good as the input information provided. Therefore, if the algorithms are not well designed or if the data is not of good quality, the results will not be very useful.

One of the main aims of Machine Learning in the area of security is to improve human analysis in all aspects, including attack detection, network analysis, and vulnerability assessment. Machine Learning algorithms are efficient and effective in protecting digital environments with regard to the continuous monitoring of network traffic, allowing the detection of malicious activities that could go unnoticed, the closed-loop detection and response systems, capable of predicting future actions based on data analysis and even considering the detection of anomalies in behavior between IoT devices.

8.5 TRENDS

A Blockchain is a protection tool considered practically impossible to be breached, which is a permanent database and shared by a chain of users. In simplified terms, the main singularity is that, in order to corrupt a common computer system, it is necessary to hack one or more servers on which it is hosted. However, to violate Blockchain technology, it would be necessary to individually invade thousands of intertwined and encrypted nodes around the planet within a few minutes. Given this difficulty brought to this system representing an unprecedented degree of invulnerability regarding digital security, especially with respect to IoT (França et al. 2020a–2020f).

Digital twins are a virtual counterpart to a real object, which means it can be a product, structure, or system. They are the digital representation of an entity. Considering that over time, digital representations of virtually all aspects of the world in which society lives will be dynamically connected with their real-world counterpart, with each other and loaded with AI-based resources to allow simulation, operation, and advanced analysis. This technology in the context of IoT projects is particularly promising in the coming years. Since this is designed with the potential to considerably improve decision making in companies, i.e., innovation is used to understand the state of the product or system, improve operations, respond to changes, and add value. Still pondering that the technology allows modeling a scalable and secure environment, breaking silos that previously hindered perceptions and providing a robust platform for building dynamic business logic, connecting assets, such as IoT devices (El Saddik 2018).

In short, it is the virtual version of a product or its production line (in the context of industry 4.0). It is a kind of digital mirror where a series of emerging technologies such as IoT, Big Data, Cloud, Analytics, and simulation software play key roles. Enabling insights to help create, optimize operations and costs, and create innovative experiences and connected solutions (Ustundag & Cevikcan 2017).

Edge Computing is essential for IoT to scale in performance, security, and user experience, describing a type of computing in which information processing and the collection and delivery of content are placed closer to the sources of that information, i.e., technology ends in the Cloud. Since the technology overcomes the challenges of connectivity and delay, bandwidth restrictions favoring distributed models, considering the use of Edge design patterns in infrastructure architectures, particularly for those with significant elements of IoT (França et al. 2020a–2020f).

Also relating that the Cloud Computing architecture depends on many links (each one is a potential point of failure) in a communication chain to move data from the physical world of IoT devices to the digital world of IT, through this factor, the Edge Computing concept is growing. Given that Edge Computing is very close to the perimeter (limit or edge of the network), optimizing the connection and increasing the response time. With technology, information first goes to a gateway, i.e., an intelligent intermediate platform performing processing, decision making, and interconnecting small networks, and then transferring the data to a Cloud architecture or even to a Fog Computing architecture (França et al. 2020a–2020f).

Thus, the sending of data that is being collected by the IoT sensors to the Cloud in a centralized processing format causes an increase in costs with communication infrastructure for data transmission. One solution is to do a pre-analysis of the data at the place where it is collected and only then carry out the transfer of the data that matters, offering a higher level of performance and complex processing at the place, minimizing risks and with significant gains for real-time analysis. The technology also allows significantly minimizes the risks of cyber-attack, considering if a server suffers a problem, few users are at risk of being without the service. Thus, with the Edge approach, it does not depend on a concentrating server. Besides, the connection close to the user allows minimizing the impacts of latency, dependence on the transmission network, and preceding the loading of content (Padilha 2018).

Adopting endpoint protection is a way to prevent IoT devices from being attacked and contaminating the corporate network, given that this type of security software allows the use of firewalls, antivirus (protection against viruses, malware, spyware on workstations) to be articulated, and servers), encryption (disk, files, and folders) and even two-point verification to ensure the best possible protection. Also, ensuring that when the device connects to the network, the server will perform a full scan, ensuring that it operates within the security rules before enabling its access (Awad et al. 2018).

Intrusion detection and prevention is a managed service for monitoring, detecting, and preventing intruders at stations and servers installed at endpoints (physical/virtual servers/stations), also correlated Host IPS (HIPS) that help identify and/or block threats that may attempt to penetrate critical assets on your network, such as IoT devices, by hiding in traffic that would normally be considered legitimate (Fraley & Cannady 2017; Awad et al. 2018).

Also relating protection against information theft (DLP, e.g., Data Loss Prevention) consisting of a system used to ensure that confidential data is not lost, stolen, accessed by unauthorized people, misused, or leaked by malicious users. DLP consists of a set of security policies and rules that can be applied with the help of specialized software to enforce blocking against possible intruders, considering it to be applicable to increase security on IoT devices (Awad et al. 2018).

Network access control (NAC) was designed to handle large corporate networks that have various types of devices (IoT) trying to connect all the time, while still pondering that this is mainly proactive, which means that they seek to block or stop attacks before they come true. Considering the main objective of these solutions is to defend the entire network perimeter, being able to reduce these risks in IoT devices by applying defined profile and access policies in several categories of devices, even including physical infrastructure and any Cloud-based system that is linked, capable of including IoT hardware in their access policies. An important function is the ability to inventory and tag each unknown piece of hardware within the network, categorizing devices into groups with limited permission and constantly monitoring activity so that devices have not been invaded (Baird et al. 2017).

8.6   CONCLUSIONS

In recent years, the task of cybernetic protection and security has become even more complicated, which is an increasingly important issue for all organizations and society due to the new classes of IoT devices and their proliferation, each of them having new surfaces of attack and contradicting the well-defined perimeter concept.

Also relating those old models of digital security are based on the concept of a well-defined perimeter and the use of tools with rules and statistical analysis and signatures. Although these models have been widely used for a long time, they have become tools of little significance and, with many limitations in the face of large-scale security breaches, created to avoid these traditional defenses. This increases the need for security development more and more as a way to protect and guarantee the protection not only of the internal information of organizations of various types but also the information of users, especially after the new data protection laws around the world.

With regard to IoT, it affects all sectors of society, with essential use for business in many vertical sectors, including health, retail, agriculture, and transportation, among many others. With respect to IoT applications and solutions, the lack of effective security is the main obstacle to successful IoT initiatives, requiring the adoption of a much higher degree of visibility of your IoT device networks that might have been necessary before.

AI focuses on the theory and development of computer systems capable of performing tasks that normally require human intelligence, such as visual perception and decision making. More precisely, Machine Learning is a subset of AI that uses algorithms to analyze data, learn from it, and then make predictions based on that learning. Unlike static algorithms, an important aspect of Machine Learning is that the machine is "trained" using large amounts of data and algorithms that guarantee the ability to continuously learn how to perform a given task.

Machine Learning is an essential ally in the digital security process, making it possible to significantly increase the digital protection of systems by preventing the actions of cybercriminals who wish to usurp or illegally manipulate data. Also, considering that AI-enabled security solutions help to reduce false alerts, increase effectiveness, and make investigations more efficient by accelerating discovery and response to sneak cyber-attacks since technologies like Machine Learning are essential to detect attacks against digital security before they can cause harm.

Machine Learning-based tools are needed to complement the current set of security tools, as the technology helps organizations to identify and mitigate the new generation of security breaches that target both old and new attack surfaces so that they are not detected by traditional defenses. To maximize the value of Machine Learning, the tool must have access to the widest possible data set (considering that in almost all instances, the network is the best data source), including data flows, data packages, histories, and alerts. Still pondering that the technology allows the system to identify what are the normal patterns and even the pattern of cyber-attacks, and when there is a deviation from the pattern, an alert is issued that can prevent the DDoS attack, for example.

The results produced by Machine Learning-based security tools are not binary, considering that the results would not be used to change a green to red alert, but indicate the likelihood of a security breach and to identify the types of small incremental changes that normally cannot be detected by traditional tools indicating that violation. Another important factor is that some systems are already capable of improving themselves, fixing possible loopholes that may arise, reducing digital security weaknesses even before human IT teams can identify, streamlining the process, and make the necessary changes.

REFERENCES

Alpaydin, E. (2020). *Introduction to Machine Learning*. MIT press.
Al-Turjman, F. (Ed.). (2019). *Artificial Intelligence in IoT*. Springer, Berlin.
Awad, A. I., Yen, N., & Fairhurst, M. (2018). *Information Security: Foundations, Technologies and Applications*. Institution of Engineering & Technology.

Babu, P. D., Pavani, C., & Naidu, C. E. (2019). Cyber security with IoT. In *2019 Fifth International Conference on Science Technology Engineering and Mathematics (ICONSTEM)* (Vol. 1, pp. 109–113). IEEE, New York.

Baird, M., Ng, B., & Seah, W. (2017, June). Wi-Fi network access control for IoT connectivity with software-defined networking. In *Proceedings of the 8th ACM on Multimedia Systems Conference* (pp. 343–348).

Charniak, E. (2019). *Introduction to Deep Learning*. The MIT Press.

Cielen, D., Meysman, A., & Ali, M. (2016). *Introducing Data Science: Big Data, Machine Learning, and More, Using Python Tools*. Manning Publications Co.

El Saddik, A. (2018). Digital twins: the convergence of multimedia technologies. *IEEE Multimedia*, 25(2), 87–92.

Flasiński, M. (2016). *Introduction to Artificial Intelligence*. Springer, Berlin.

Fraley, J. B., & Cannady, J. (2017, March). The promise of machine learning in cybersecurity. In *SoutheastCon 2017* (pp. 1–6). IEEE, New York.

França, R. P., Iano, Y., Borges, A. C., Monteiro, R. A., & Estrela, V. V. (2020a). A proposal based on discrete events for improvement of the transmission channels in cloud environments and big data. In *Big Data, IoT, and Machine Learning: Tools and Applications*, 185.

França, R. P., Iano, Y., Monteiro, A. C. B., & Arthur, R. (2020b). A review on the technological and literary background of multimedia compression. *Handbook of Research on Multimedia Cyber Security* (pp. 1–20). IGI Global.

França, R. P., Iano, Y., Monteiro, A. C. B., & Arthur, R. (2020c). Improvement of the transmission of information for ICT techniques through CBEDE methodology. *Utilizing Educational Data Mining Techniques for Improved Learning: Emerging Research and Opportunities* (pp. 13–34). IGI Global.

França, R. P., Iano, Y., Monteiro, A. C. B., & Arthur, R. (2020e). Potential proposal to improve data transmission in healthcare systems. *Deep Learning Techniques for Biomedical and Health Informatics* (pp. 267–283). Academic Press.

França, R. P., Iano, Y., Monteiro, A. C. B., & Arthur, R. (2020f). Potential proposal to improve data transmission in healthcare systems. In *Deep Learning Techniques for Biomedical and Health Informatics* (pp. 267–283). Academic Press.

França, R. P., Iano, Y., Monteiro, A. C. B., & Arthur, R. (2021). Better transmission of information focused on green computing through data transmission channels in cloud environments with Rayleigh fading. In *Green Computing in Smart Cities: Simulation and Techniques* (pp. 71–93). Springer, Cham.

França, R. P., Iano, Y., Monteiro, A. C. B., Arthur, R., & Estrela, V. V. (2020d). A proposal based on discrete events for improvement of the transmission channels in cloud environments and big data. *Big Data, IoT, and Machine Learning: Tools and Applications*, 185.

François-Lavet, V., Henderson, P., Islam, R., Bellemare, M. G., & Pineau, J. (2018). An introduction to deep reinforcement learning. arXiv preprint arXiv:1811.12560.

Goodfellow, I., Bengio, Y., Courville, A., & Bengio, Y. (2016). *Deep learning* (Vol. 1). MIT Press, Cambridge.

Handa, A., Sharma, A., & Shukla, S. K. (2019). Machine learning in cybersecurity: a review. *Wiley Interdisciplinary Reviews: Data Mining and Knowledge Discovery*, 9(4), e1306.

Hsu, K., Levine, S., & Finn, C. (2018). Unsupervised learning via meta-learning. arXiv preprint arXiv:1810.02334.

Hutter, F., Kotthoff, L., & Vanschoren, J. (2019). *Automated machine learning: methods, systems, challenges* (p. 219). Springer Nature, Cham.

Jackson, P. C. (2019). *Introduction to artificial intelligence*. Courier Dover Publications.

Khan, M. A., & Salah, K. (2018). IoT security: review, blockchain solutions, and open challenges. *Future Generation Computer Systems*, 82, 395–411.

Liu, G., Quan, W., Cheng, N., Zhang, H., & Yu, S. (2019). Efficient DDoS attacks mitigation for stateful forwarding in Internet of Things. *Journal of Network and Computer Applications*, 130, 1–13.

Meneghello, F., Calore, M., Zucchetto, D., Polese, M., & Zanella, A. (2019). IoT: Internet of Threats? A survey of practical security vulnerabilities in real IoT devices. *IEEE Internet of Things Journal*, 6(5), 8182–8201.

Miloslavskaya, N., & Tolstoy, A. (2019). Internet of Things: information security challenges and solutions. *Cluster Computing*, 22(1), 103–119.

Mohri, M., Rostamizadeh, A., & Talwalkar, A. (2018). *Foundations of machine learning*. MIT Press.

Monteiro, A. C. B., Iano, Y., França, R. P., & Arthur, R. (2020). Development of a laboratory medical algorithm for simultaneous detection and counting of erythrocytes and leukocytes in digital images of a blood smear. In *Deep Learning Techniques for Biomedical and Health Informatics* (pp. 165–186). Academic Press.

Mylrea, M., Gourisetti, S. N. G., & Nicholls, A. (2017). An introduction to buildings' cybersecurity framework. In *The 2017 IEEE Symposium Series on Computational Intelligence (SSCI)* (pp. 1–7). IEEE, New York.

Neapolitan, R. E., & Jiang, X. (2018). *Artificial Intelligence: With an Introduction to Machine Learning*. CRC Press.

Padilha, R. F. (2018). Proposta de um método complementar de compressão de dados por meio da metodologia de eventos discretos aplicada em um baixo nível de abstração (Proposal of a complementary method of data compression by discrete event methodology applied at a low level of abstraction). http://repositorio.unicamp.br/jspui/handle/REPOSIP/331342

Raschka, S., & Mirajalili, V. (2019). *Python Machine Learning* (Vol. 1, pp. 1–18). Packt Publishing.

Rebala, G., Ravi, A., & Churiwala, S. (2019). Machine learning definition and basics. *An Introduction to Machine Learning* (pp. 1–17). Springer, Cham.

Rojko, A. (2017). Industry 4.0 concept: background and overview. *International Journal of Interactive Mobile Technologies (iJIM)*, 11(5), 77–90.

Roopak, M., Tian, G. Y., & Chambers, J. (2019, January). Deep learning models for cybersecurity in IoT networks. In *2019 IEEE 9th Annual Computing and Communication Workshop and Conference (CCWC)* (pp. 0452–0457). IEEE, New York.

Shovic, J. C. (2016). Introduction to IoT. *Raspberry Pi IoT Projects* (pp. 1–8). Apress, Berkeley, CA.

Simeone, O. (2017). A brief introduction to machine learning for engineers. arXiv preprint arXiv:1709.02840.

Stamp, M. (2017). *Introduction to Machine Learning with Applications in Information Security*. CRC Press.

Sterne, J. (2017). *Artificial Intelligence for Marketing: Practical Applications*. John Wiley & Sons.

Strong, A. I. (2016). Applications of artificial intelligence & associated technologies. *Science* 5(6).

Thames, L., & Schaefer, D. (2017). *Cybersecurity for industry 4.0*. Springer, Heidelberg.

Tran, C., & Misra, S. (2019). The technical foundations of IoT. *IEEE Wireless Communications*, 26(3), 8–8.

Tweneboah-Koduah, S., Skouby, K. E., & Tadayoni, R. (2017). Cybersecurity threats to IoT applications and service domains. *Wireless Personal Communications*, 95(1), 169–185.

Ullah, Z., Al-Turjman, F., Mostarda, L., & Gagliardi, R. (2020). Applications of Artificial Intelligence and Machine learning in smart cities. *Computer Communications*.

Ustundag, A., & Cevikcan, E. (2017). *Industry 4.0: Managing the Digital Transformation*. Springer, Berlin.

Zhou, L., Pan, S., Wang, J., & Vasilakos, A. V. (2017). Machine learning on big data: opportunities and challenges. *Neurocomputing*, 237, 350–361.

9 Qualitative and Quantitative Evaluation of Encryption Algorithms

Rakesh Garg, Supriya Raheja, and Durgesh Pandey

Amity University, Noida

CONTENTS

9.1 INTRODUCTION

Encryption has always been a matter of concern for centuries, and the systems have evolved to suit the needs of the context. We evolved from using simple Morse code over telephone lines to using sophisticated encryption algorithms backed by mathematical and scientific proofs. But as the systems evolved, it got difficult to decide the context of each algorithm. Which one to use and which one to skip became difficult over time. It also became evident that, as the cryptanalysis systems got better, so did the anti-cryptologists. With every new algorithm introduced in the industry, an older one got bypassed or was broken into. This caused a ripple in the cryptanalytic

DOI: 10.1201/9781003140023-9

society, which did hit every algorithm. From there began a battle between those trying to protect and those trying to break through (for good or for intentionally bad deeds). It still continues to this time and will continue until either we come to an agreement or we create an unbreakable cryptographic algorithm. As evident, either of them seems impossible to happen. With the evolution of technology came the era of digitization, and all the ciphers and algorithms were now performed by computers.

In this fast-paced world, where everything is digitalized, and data is stored over the internet, computation is done over the cloud, and people have more passwords online than their safe in their homes, it becomes important to keep the data online safe and secure. As the transmission of this data is over networks, there are clear chances of someone exploiting the loopholes in the network and gaining access to the sensitive data. Privacy is an important aspect of the internet, and we do not need it to be compromised at any time. As more and more algorithms come into existence, it becomes more and more difficult to determine all the possible algorithms that could serve the purpose. Here we try to discuss the ways in which we could associate a numerical value with each algorithm which would represent that algorithm in a comparative analysis for the best. But it is also important to note that there are qualitative aspects associated with each of the algorithms too. Considering only the numbers might not provide us with the most efficient algorithms, but considering the non-numerical factors of the algorithm, we may get a very good idea of the position of the algorithm in the list of comparisons. Hardware-related discussion is beyond the scope of this paper, and so we stick to the general-purpose software side.

Before going into the numerical aspects and figurative facts, it is necessary to define some parameters on which the algorithms will be assessed and hence evaluated. For the sake of better understanding and common ground, we consider the time of execution, memory acquired, block size, key size, and code difficulty and breaches. These parameters are common to all the algorithms and hence form a solid ground for analysis and evaluation. The coding language used to write, test, evaluate and analyze these algorithms is JAVA.

Further, Section II describes the algorithms that are used in the study and their qualitative and quantitative aspects. Each algorithm is described in brief and used in a particular form (for those available in multiple forms). Section III describes the ranking formula derived from the relations established between factors and the final expression that results from a logical combination of these expressions. Section IV takes on the quantitative comparison of the algorithms and associates every algorithm with a number giving its relative rank. Section V analyses the rank from a qualitative perspective and brings out the possible qualitative loopholes or downsides of some algorithms or vice versa. Section VI compares the results of this study from that of earlier comparative studies of a similar type. It is worth mentioning that the results of the study under comparison must contain the parameters under consideration in this study to have an unbiased result comparison. Section VII gives the conclusions from this study and the edge cases of the algorithm. Section VIII discusses the limitations of this study and its findings and also the future work which is possible on this study or on the basis of the results of this study.

9.2   ENCRYPTION

Encryption is the process of converting data or information from one form to the other so as to protect it from direct interception and maintain data security. Encryption is a process which can be applied to any form of data or information, so it can be applied to any form of data. From files to raw text information, everything can be encrypted. To achieve encryption, we just need to transform the data in such a way that the result is rendered incomprehensible (probably).

Encryption has been evolving ever since it came up as a subject of study and an important aspect of data security. Starting as ciphers dating back to Roman times, where Julius Caesar used them to transform instant military information by shifting the characters of the words by some units. The **Caesar Cipher** is one of the simplest ciphers out there, which can be cracked easily just with time and some effort. It is a type of shifting cipher which uses a wheel with characters and shifts them by some units. Consider the character 'A,' which has an ordinal value of '0'. Then the inner wheel will be fixed to the board, and the outer wheel will have the same markings. To create the cipher, we rotate the wheel in one direction and note the number that falls on the character 'A' of the inner wheel. This will be our key for decryption, and the rest of the process is just search-and-replace. We find the character on the inner wheel and replace it in the output with the corresponding character from the outer wheel.

This shows that we were consistently inclined toward developing mechanisms and methods which would facilitate secure communication of information.

The most commonly used algorithms include MD5, SHA256, HMAC512, AES256. These are strong algorithms, but what makes them so strong? What does it take to decrypt them by brute force? How much time would it take to brute force the last key of such an algorithm? Let's answer these questions one by one, starting with the types of encryption algorithms that are available to us currently.

We can broadly classify encryption algorithms into two categories. The **symmetric** encryption systems and the **asymmetric** encryption systems. Let's talk about each of those.

9.2.1   SYMMETRIC ENCRYPTION SYSTEMS

Symmetric Systems use a single key for both encryption and decryption. That means one who has the key has control of the information. The key is used to encrypt (transform the input in some way which is, logically and mathematically, related to the key) the data, and the data and key are sent separately to the consumer or the end person. Advanced Encryption Standard (AES), Data Encryption Standard (DES), Blowfish, RC5, RC6, IDEA are some symmetric encryption systems. We can perform the same encryption multiple times to obtain a different set of transformations, but we need to remember the order of the keys used. That is necessary because while decrypting the information, the reverse order needs to be followed for successful decryption.

These types of systems are very susceptible to "Man in the Middle" type attacks. This involves someone sitting between you and your other communicating device. Consider this in the context of a scenario. You are in a café, and you are using their public Wi-Fi to send some "encrypted" information to your friend. Suppose someone is able enough to sniff the packets from the "public" network does his trick. He might easily obtain the key and the data and sit between you and your friend and watch all the data being exchanged. This is called "Man In The Middle" attacks [MITM attacks]. This is harmful because it does not matter who exposes the key. Once the key is found, the data can easily be decrypted from both sides without any additional effort.

This is considered to be one of the weak aspects of symmetric encryption systems, and so asymmetric systems were developed to overcome this issue.

9.2.2 ASYMMETRIC ENCRYPTION SYSTEMS

Asymmetric Encryption systems use multiple and different keys for encryption and decryption. Because the keys are different and are not exposed at the same time, anyone with one key can only perform the operation supported by the key (in some regards). This allows us to limit the use of keys to a context and so eliminate the MITM attacks to a great extent.

The asymmetric part of the system is for the keys. We use one key for encryption and the other for decryption. And we ourselves generate those keys, which means if one key is compromised, we can generate a new "pair" of keys and restart the process without having to troubleshoot all the systems.

Let us get this in a context so as to better understand what we are working with here. Consider two people A and B. Both have their mailboxes which are available to the public, and anyone can drop them a mail anytime. But only the owner of the mailbox has the key to open the box and check the mails. This way, even if someone finds out their mailboxes, all they can do is drop them a mail. With regards to the encryption system, each node (A and B here) generates its own public-private key pair. The public key is shared with anyone who wishes to communicate with them, and this key is used to encrypt the data and send it to the owner of the key. Then the owner uses their "private" key to decrypt the data and get the underlying information. MITM is not effective here because even though one key is being shared over the network, the public key is always used to encrypt data and so it cannot be used to decrypt the same data, while the private key is kept secret and used for decryption and NEVER communicated over any network.

These algorithms are more complicated but are stronger. The key pairs can be generated anytime, and so for any compromised key, we can use new keys and better internal security to avoid any compromises.

RSA is one of the most commonly used asymmetric encryption systems. SSL, which is used to secure websites, also uses a similar algorithm wherein the certificate uses a keypair to verify the signature of the certificate. Any tampering with the certificate will result in a void in security and so is always verified before the site is loaded completely into the browser.

9.2.3 WHAT MAKES THEM STRONG?

Thinking naïvely, the first thing in one's mind after reading till here would probably be, "The key is what makes the system stronger." Well, you are pretty much right.

The strength of the key is a big factor and for now, let us see how the bit size of the key can affect the cryptographer's efforts in brute-forcing the actual key.

Let us simply consider a human or a machine working at the speed of an average human. Let us give it a speed of about 1 computation per second, where computation refers to making the key, trying it with the encrypted information, and validating the output. For a key of size 10 bits, there are 2^{10} possible keys. Why is that? There are 10 bits to be filled, and each bit can take any one of the binary values 0 or 1. So for 1 bit, we have two choices, and for n bits, we have 2^n choices.

If the machine tries all the combinations, it would take about 17 minutes to work out the key. Not so scary, huh! Okay, now let's make the key a little stronger. Let's take a key of size 25 bits. How long would that take...? Just a little over a YEAR. About 388 days to be close. That is a fairly long time. What about a key size of 50? That would take our machine about 3,56,77,615.3 years. Yes, that is about 3.5 billion years. Now that is a very long time.

But our machines are faster, you might ask. Yes, they are, but our keys are bigger as well, we use 128-bit, 196-bit, 256-bit, 512-bit keys to secure our systems, and along with that, we use asymmetric systems for additional security.

One drawback, though, is the high CPU demand for asymmetric systems and high resource usage. Thus, they are primarily used to encrypt much secure information where encryption and decryption do not occur very often.

While our systems evolved, our machines and technologies evolved, so did malicious minds. They are ready to act on any new technology, finding weak points and exploiting them for "god knows what" reasons. Hence our data security is one of the important concerns in modern times, and in the time to come, we will have more sophisticated security systems and protocols.

9.3   ALGORITHMS UNDER CONSIDERATION

The following algorithms are used in the process of evaluation and comparison of performance. The security aspects of the Algorithms have been considered only after the performance and cost of operation are found feasible. Cost of operation can be defined as the amount of time taken to provide output to the console/client. Higher cost of operation means more time taken to provide the same amount of output with respect to other Algorithms (considering defined factors to be constant).

- AES
- 3-DES
- RSA
- TwoFish
- BlowFish

9.3.1 ADVANCED ENCRYPTION STANDARD

The Rijndael algorithm is the new AES approved by the US National Institute of Standards and Technology (NIST). With this algorithm supporting significantly larger key sizes than DES supports, NIST believes that the AES has the potential of remaining secure for the next few decades [1].

As stated in the FIPS PUB 197: AES. This standard specifies the Rijndael algorithm, a symmetric block cipher that can process data blocks of 128 bits, using cipher keys with lengths of 128, 192, and 256 bits [2, 3]. Rijndael was designed to handle additional block sizes and key lengths. However, they are not adopted in this standard. The adoption of this standard was a need as the DES algorithm failed to serve the security purpose of then-existing systems. The most common application of the DES system was the banking system, where transactions and vital information were transmitted over a network.

It processes the input data in 128-bit blocks and a key length of 128, 192, and 256 bits. It takes 9, 11, or 13 rounds (respectively, for each key size), with each round performing four fundamental steps on the blocks. It is an iterative cipher instead of the common Feistel Cipher.

9.3.2   3-DES – TRIPLE DATA ENCRYPTION STANDARD

Before the proposal of the AES standard, 3-DES was proposed as the successor of the old DES system. While the DES used 56-bit key size and 64-bit block size, 3-DES used 64-bit block size (While the block size of the newer version was increased to 64-bits, the last 8 bits were reserved as signing bits, and so the apparent block size of the algorithm came out to be 56-bits only) with a larger key size of 168, 112 or 56-bits. The key size of 3DES is determined by the no of distinct keys used. If three distinct keys are used, then the key size will be 168 bits (3 x 56 bits). Similarly, for two distinct keys, the key size will be 112 bits (2 x 56 bits). And so, it will be for a single key, which is 56 bits, taken from DES. It takes three keys for best operation and hence was comparatively secure and more reliable. But it was necessary to assess the performance of this algorithm and determine its position in the efficiency table.

9.3.3   RSA – RIVEST-SHAMIR-ALDELMAN

This is an asymmetric encryption algorithm, meaning it uses different keys for encryption and decryption. While the encryption key is public, the decryption key is kept secret. A key of this system is typically 1024 bits or 4096 bits. Being an asymmetric algorithm, it does not perform multiple rounds and so might be considered faster. It is a mathematical algorithm involving prime and co-prime numbers, bit operations, and bit to digit or string conversions and vice versa [3, 4]. The algorithm is a stream-based cryptography system, and so there is no block size for this. Hence, in calculations, we consider the (apparent) block size to be equal to the key size.

9.3.4   TWOFISH

TwoFish is a derivative of the older Blowfish algorithm, which accepts a variable key length up to 256 bits (typically 128, 192, or 256 bit). The block size for this derived block cipher is 128 bit. It was one of the finalists for the candidature of the AES algorithm discussed before. Distinct features of this algorithm include pre-computed, key-dependent S-boxes and a more complex key schedule. The key scheduler was responsible for generating the round keys for each round. It was slower than Rijndael for 128-bit keys but was comparatively faster for 256-bit keys, but after standardization of Rijndael, it became much slower for CPUs that supported the AES instruction set [5, 6].

9.3.5 BLOWFISH

It is a symmetric key algorithm, meaning it uses the same secret key to both encrypt and decrypts the message. It uses a block size of 64 bits, with key sizes ranging from 32 bits to 448 bits. It also uses the standard 16 rounds from the Feistel Cipher, as in the DES algorithm, to encrypt the data. The key scheduler is the same as the DES algorithm. It uses the large key-dependent S-boxes. Schneier developed this algorithm as an alternative to the aging DES algorithm and kept it as an unpatented algorithm, free to use for all across the countries.

9.4   RANKING FORMULA

The process adapted for evaluation is simple and easy. All the algorithms are implemented using JAVA for clarity in execution and evaluation. Because the inputs are the same for all the algorithms, and hence we eliminate the input variable from the evaluation. Next, as we also see that the execution depends on system architecture, so we execute, analyze and evaluate all the algorithms on the same computer system under the same conditions (including system temperature and clock speed). Hence, we eliminate one more variable from the equation. This brings us down to the equation that simply says:

> The execution time, block size, key size, input size, peak memory occupied and the code breaches, are the only solid grounds of evaluation of the algorithms, after eliminating all the possible variables from the system.

After studying algorithms, the parameters that were most evidently impacting the efficiency of the algorithm are:

- Time of execution (written as t, in sec)
- Key size (written as k)
- Block Size (written as b)
- Size of the input data (written as N)
- Passes that the algorithm makes during execution (written as p)

All these had individual relation with the efficiency of the algorithm. As we cannot directly measure the efficiency of any algorithm, we need some kind of standard to compare the score of the algorithms. Hence, we observe obvious relations between the efficiency of an algorithm and each of the above independent factors.

It can be noted that for any algorithm, time of execution or cost of operation gives an indication of its efficiency. Design and Analysis of Algorithms is a subject under which we study the fact that the execution time of an algorithm is inversely proportional to its efficiency. The less time an algorithm takes for its computation, the faster it is, as mentioned in equation (9.1),

$$efficiency \propto \frac{1}{t} \qquad (9.1)$$

The more no. of passes that the algorithm will make in a given time, the more efficient and secure it will be. Thus, we can say that

$$efficiency \propto p \qquad (9.2)$$

In cases, the algorithm is based on a block cipher, the value of p is calculated as the ratio of key size (in bits) to that of the block size (in bits). Mathematically, we can say

$$p = \frac{\left[keysize\left(int\ bits \right) \right]}{\left[blocksize\left(in\ bits \right) \right]} \qquad (9.3)$$

Combining the equations (9.1) and (9.2), we can say

$$efficiency \propto \frac{p}{t} \qquad (9.4)$$

For stream ciphers, where the block size is not available or cannot be determined, the value of p is considered to be 1, i.e., the block size is equal to the key size. This is because the stream of data is most probably processed in chunks of key size. Hence, we can assume that the stream of data is processed in blocks, each size equal to the key size of the cipher.

Also, we know from the Design and Analysis of Algorithms that time increases as N increases. This is because of the fact that most encryption Algos involve loops or have time complexity of $O(n2)$ or $O\ (n\ \log n)$. Hence, we define a new parameter θ which will be our input factor, which will compensate for the exponential growth of time complexity.

$$\theta = \log 10\left(N \right).$$

Now, we find out a numerical score by combining the above equations, and we call this β-score. Therefore,

$$\beta\ score = \ln\left[\frac{\theta\ p}{t} \right]$$

From the above equations and explanation, it is evidently clear that all the stated factors are present in this equation, and it satisfies all the above equations. We define the dimension of input-factor (θ) as s. This makes the equation dimensionally correct and mathematically correct.

Also, to note that, in the final equation, we end up taking double logarithms. The first one is for input-factor θ, on base 10, and the second for the equation itself, on base e. The first one compensates for the fact that; the Input size grows exponentially, and we can shrink it to smaller values using the log base 10. The second log makes the ultimate score smaller and into a considerable range, including negative values. The reason for the natural log is that it gives a value of 1 only when the input equals e, which is very rare, and so we get a decimal value every time we run the equation through the parameters. This also gives us a precise difference between the algorithms' scores.

9.5 QUANTITATIVE OBSERVATIONS

We observe the data from the algorithms we considered in this paper. Before we begin with experimentation, we need to define the machine that we used to observe the data. The specifications of the machine are aforementioned.

- Intel core i5-9300H @ 2.40 GHz (8 cores)
- 64-bit OS architecture
- 8192 MB primary memory (RAM)
- 7984 MB graphic memory (GPU)
- CPU temp. ~38–40°C (Maintained)

All algorithms are evaluated on the above-mentioned specifications. The values obtained may vary from machine to machine but will provide a clear idea of the efficiency of any algorithm at given parameters. Table 9.1 shows data we obtained on analysis of five common encryption algorithms.

Now, using the above formula, we calculate the value of β-score for each of the algorithms above, using the observed data.

TABLE 9.1
Parameter Values of Five Common Algorithms

Algorithm	Time (T)	Key Size (K)	Block Size (B)	P	Input Size (N)	Input Factor (θ)
AES	1.5625 sec	256	128	2	65,536,000	7.8164799
RSA	1.921875 sec	1024	1024	1	65,536,000	7.8164799
3-DES	7.6875 sec	168	54	3.111	65,536,000	7.8164799
TWOFISH	0.046875 sec	64	128	0.5	65,536,000	7.8164799
BLOWFISH	10.28125 sec	64	256	0.25	65,536,000	7.8164799

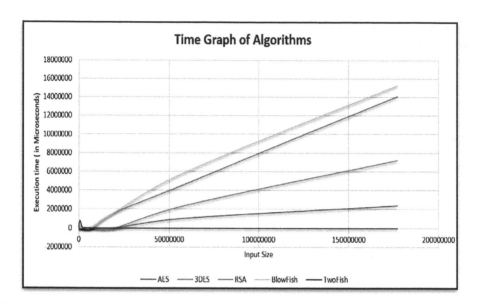

FIGURE 9.1   Time graph for all algorithms.

9.6   QUALITATIVE ANALYSIS VS. NUMBERS

We can fairly observe from the above data observations that β-score (β in short) of 3DES and RSA algorithms are very close. This gives a fair idea that we have a choice here. We can go with either of them, as per our contextual needs. Observing the time of execution from Figure 9.1, of both alone, one may conclude that RSA must be more efficient as it is faster, but on the other hand, 3DES has a larger key size and smaller block size, and so it takes time to operate on the keys w.r.t to the blocks. As the time of computation is in secs (seconds), we may neglect the difference with respect to the input size.

If the requirement is of an algorithm, which encrypts fairly fast and where security is a secondary concern, AES might be the best choice. While on the other hand, if one needs a more secure algorithm in similar computational time, then RSA will be the best competitor of AES here.

Also, the data shows that the TwoFish algorithm has the best β score (4.42335) (Table 9.2). We also see that for the same input size, TwoFish takes 97% less time than

---

TABLE 9.2

Ranking Algorithms Based on β-Score

Algorithm	β-Score	Rank
TWOFISH	4.423357931	1
AES	2.303094394	2
RSA	1.402933044	3
3-DES	1.151582901	4
BLOWFISH	−1.660381893	5

AES to perform the same operation. Being a symmetric algorithm, the security can be compromised. But if the context involves using an encryption algorithm to transfer data in an intranet, then this will be the fastest algorithm to serve the purpose.

Notice that the Blowfish algorithm has a negative score. This implies that the algorithm is slower. The reason for a negative value is purely mathematical.

When $\left\lceil \dfrac{\theta p}{t} \right\rceil < 1$ we get a negative score

A negative score does not mean that an algorithm is bad, it is just a number associated with the algorithm, and it indicates its position. Also, it does not consider the qualitative facts of the algorithm like, it is the extent of use and its security. The execution time graph shows the somewhat exponential growth of time with respect to the input size. Starting from a few microseconds to about 15 seconds, these algorithms take a different amount of time to perform the same operation. As discussed earlier, TwoFish works faster on workspaces with higher internal memory (RAM). The test environment stated before had 8 gigabytes of memory, enough for a simple TwoFish algorithm to reach its maximum potential and perform the best. Yet, at smaller inputs, it does not perform well and displays a high time of execution. BlowFish takes longer to process and provide output.

9.7 COMPOSITION OF RESULTS

The paper compared the results of the presented work with the results of Performance Analysis of Encryption Algorithm in Cloud Computing. As there is no disclosure of the key size of the algorithms in the paper, we consider that the algorithms are stream-based ciphers (CBC ciphers), and hence their key size and block size will be equal. Table 9.3 shows the comparison results. For evaluation, we have considered the input size and execution time from the referred paper and then proceed with our working formula.

From our score, we can conclude that the Blowfish algorithm has the best score out of all algorithms. This is consistent with their conclusion and results, which say that when taking it to the cloud, AES and Blowfish are the most preferable among their options. Our score provides a much better view and mathematical proof of why it is preferable. The difference of score between AES and Blowfish is almost 0.2259. It might seem large, but with respect to DESede, which has a much higher deviation from their scores, this is a preferable option. Also, as they stated, AES has an edge of speed over Blowfish when considering larger input sizes. On the cloud, managing

TABLE 9.3
Validating Older Results Using Our Formula

Algorithm	Input Size	Time	β-Score	Rank
AES	57344 (56 KB)	3.75 ms	0.238174	2
BlowFish	57344 (56 KB)	3 ms	0.4641317	1
DESede	57344 (56 KB)	14.5 ms	−1.114219	3

resources is a big deal, and do the size of the data matters a lot, and the faster we can process it, the better it is for the algorithm.

It is also important to note that in Table 9.2, Blowfish is the worst performer, and in Table 9.3, Blowfish is the best performer. The current environment favors AES and Blowfish lags out. It might be due to the internals of the Blowfish, which make it slower but, on the cloud, Blowfish has the best performance. Also, the environment of the earlier study was different, so it is not valid to make a comparison between the results of Tables 9.2 and 9.3. Table 9.3 validates our formula using the data and results from earlier studies only.

Also, from Performance Analysis of AES and TwoFish Encryption Schemes [7, 8], the performance of TwoFish increases with an increase in RAM (or the buffer storage provided to the algorithm), and for larger input size, TwoFish performs better than AES. As visible from the table, AES has a lower score (~2× slower) than TwoFish. Thus, the environment cannot be neglected, but in the same environment, the above data holds valid. Also, from the results of [9, 10], we can conclude that considering their environment and platforms, the results obtained closely relate to the results obtained here. Also, the formula holds valid for results as old as from 1999. The values are obtained on the target system, which will be the host for the systems, so we can say that the formula helps us identify the most probable candidate from the list of possibilities. We can also isolate a handful of algorithms from a pool of all the algorithms for closer and more robust inspection.

9.8 CONCLUSIONS

In this study, we tried to analyze the various parameters that affected the performance of an encryption algorithm and tried to establish a mathematical relation between them. We also tried to associate every algorithm under study with a number which could represent its performance and provide us a mathematical approach to determining the better of the few.

This formula is very useful for comparative analysis of encryption algorithms. Given the mentioned parameters, one may easily bring out a numerical representation associated with that algorithm, which would make the comparison much easier and understandable. Algorithms and Programs need more and more optimization over time to meet the industry needs, and with little or no changes, this formula might fit into their context as well. Evidently, the numbers will indicate the speed of the operation and also its efficiency as a matter of fact.

The β-score = 0 will only be possible when either of

- Input size (N) is 1, which will imply that all the algorithms are equally likely to be used for the encryption. Also, it is evidently clear that a single character/ byte can be encrypted instantly by any cipher of any type.
- $p = 0$, which will imply that either the key size is 0 or the block size is infinity. Either of them is not possible, and so this condition is never true. This is because even for the stream-based ciphers, the data is processed in buffers, and hence the value cannot be infinite or zero. Key size can also not be zero as

every algorithm considered needs a key to perform its actions, and so this is a required parameter for the algorithms and can never be null or zero.

- Time of execution (t) is infinity, which will imply that the algorithm works for an un-deterministic period of time, and evidently, this condition means that the algorithm is broken and does not work properly. Hence, for a working algorithm, this condition is also always false, as the time of execution is always finite.

Considering all the above facts together, we see that only the first condition will give a β-score of 0, which will imply the situation as stated.

We cannot neglect the environment (the machine used and the resources available) of execution, but for a given environment, without performing heavy computations, we can perform smaller computations, and with the obtained data, we may use the above-stated formula to conclude the most probable candidates for the purpose.

9.9   LIMITATIONS AND FUTURE WORK

As stated before, the environment plays an important role in the analysis of the systems. Hence the analysis is, as of now, platform and environment-dependent. There are various factors which are to be considered in this mathematical equation to rectify the loopholes in it. Some obvious factors include RAM size, GPU size, CPU frequency, and temperature of the processor (if it is found to directly impact the efficiency of the system). The formula also does not consider the strength of the cryptosystem, it only operates on the performance impacting parameters, and so is more like a performance assessment and ranking system.

The data and operations here in this paper are carried on a local system with little or no network traffic interaction. As a result, the data provided in this paper is solely device-dependent and local machine-derived [11–13].

As the current use cases suggest, these algorithms are now being ported to the cloud for safer and more secure data transmission and storage. This paper does not deal with the cloud infrastructure of the algorithms, and so there needs to be a separate dedicated study on the cloud-based or network-based cryptography. Considering the network speed and bandwidth and other similar factors.

Our future work may include the adoption of various variants of the algorithms for comparison and strengthen the validity of the above formula by providing more solid grounds and data for the validation. We may also work on the limitations of this formula. We might develop a more robust equation from the current one, which will be platform-independent or would consider the limiting factors in the equation, making them parameters to the formula.

It is also important to note the fact that, as the technology develops, most of the ciphers are countered and breached. Multiple advanced ciphers and safe locks are breached and broken. While considering the use cases, one should also accept the open breach cases and the level of security that the particular algorithm or process provides. Considering the context of use and the ease of breaking the cipher, one can make more effective decisions and save the information that needs protection.

REFERENCES

1. Jamil, T. (2004). The Rijndael algorithm. *IEEE Potentials*, 23(2), 36–38.
2. Pasham, V., & Trimberger, S. (2001). *High-speed DES and triple DES encryptor/ decryptor. Xilinx Application Notes*.
3. Rivest, R. L., Shamir, A., & Adleman, L. (1983). A method for obtaining digital signatures and public-key cryptosystems. *Communications of the ACM*, 26(1), 96–99.
4. Calderbank, M. (2007). *The RSA cryptosystem: History, algorithm, primes*. Chicago: math.uchicago.edu.
5. Moriai, S., & Yin, Y. L. (2000). *Cryptanalysis of Twofish (II)*.
6. Schneier, B., & Whiting, D. (2000, April). A performance comparison of the five AES finalists. In *AES Candidate Conference* (pp. 123–135).
7. Mewada, S., Shrivastava, A., Sharma, P., Purohit, N., & Gautam, S. S. (2015). Performance analysis of encryption algorithm in cloud computing. *International Journal of Computer Sciences and Engineering*, *3*, 83–89.
8. Rizvi, S. A. M., Hussain, S. Z., & Wadhwa, N. (2011, June). Performance analysis of AES and Two Fish encryption schemes. In *2011 International Conference on Communication Systems and Network Technologies* (pp. 76–79). IEEE, New York.
9. Hercigonja, Z. (2016). Comparative analysis of cryptographic algorithms. *International Journal of Digital Technology & Economy*, *1*(2), 127–134.
10. Nadeem, A., & Javed, M. Y. (2005, August). A performance comparison of data encryption algorithms. In *2005 International Conference on Information and Communication Technologies* (pp. 84–89). IEEE, New York.
11. Barker, E., & Mouha, N. (2017). *Recommendation for the triple data encryption algorithm (TDEA) block cipher (No. NIST Special Publication (SP) 800-67 Rev. 2 (Draft))*. National Institute of Standards and Technology, Maryland.
12. Sajay, K. R., Babu, S. S., & Vijayalakshmi, Y. (2019). Enhancing the security of cloud data using hybrid encryption algorithm. *Journal of Ambient Intelligence and Humanized Computing*, 1–10.
13. Murtaza, A., Pirzada, S. J. H., & Jianwei, L. (2019, January). A new symmetric key encryption algorithm with higher performance. In *2019 2nd International Conference on Computing, Mathematics and Engineering Technologies (iCoMET)* (pp. 1–7). IEEE, New York.

10 Analysis and Investigation of Advanced Malware Forensics

P. S. Apirajitha

Department of Information science and technology,
CEG campus, Anna University, Chennai, India

S. Punitha

Karunya Institute of Technology and Sciences,
Coimbatore, India

Stephan Thompson

M. S. Ramaiah University of Applied Sciences,
Bangalore, India

CONTENTS

DOI: 10.1201/9781003140023-10

10.1   INTRODUCTION TO MALWARE

10.1.1   DEFINITION

Malware is a general term for all sorts of malicious software. Within the context of system security, it means that "software that is employed with the aim of trying to breach a system's security policy with relevancy confidentiality, integrity, and availability" [1].

It is a piece of software meant to cause mutilation to the structure or setup. This has the litheness to pass, bring a device to its knees, and might cause the annihilation of the system.

10.2   MALWARE ANALYSIS

Malware investigation is considered or prepared to decide a given malware test's usefulness, root, and potential effect and extricate its data. The malware examination is to consider a program's behavior and confirm if it has noxious usefulness. If it is analyzed and found to be assaulted, then its classification is essential. Malware analysis is about using different tools and techniques to extract various pieces of information.

10.2.1   TYPES OF EXPLORATION

1. ***Frozen Enquiry:*** This is the method of preventing a system from executing or running it. It is used to disentangle metadata from malware as headers and strings. Inactive malware examination is performed by looking at the computer program code of malware [2]. Usually, this will help to better understand the malware's capacities, whereas when performing an inactive malware investigation, an antivirus program will run on the malware. Records such as shell scripts will be inspected. Most likely, decompiling malware ought to be performed using programs. Examples include debuggers, dissemblers, and decompiles. After this, the IT group will be able to see the source code of malware capacities.

2. ***Dynamic Analysis***: It may be a way of executing malware and analyzing its functionality and behavior [2]. Active malware examination may be a fast strategy of malware examination. When performing an active malware investigation, see how the malware acts. Records must be exchanged using a read-only medium. There are changes within the framework that ought to raise caution. It incorporates records that have been modified or included. Check for

modern administrations that have been introduced. On the off chance that any framework settings have been balanced and modern forms are running. This would join the DNS server settings of the workstation, which have been changed. Systematic activities will analyze the behavior of the framework.

3. *Code Analysis*: This is the process of analyzing and gathering code. It is the combination of an inactive and dynamic investigation.

4. *Behavioral Analysis*: This is the method of analyzing and observing the malware after execution.

10.2.2 PLATFORMS OF MALWARE STUDY

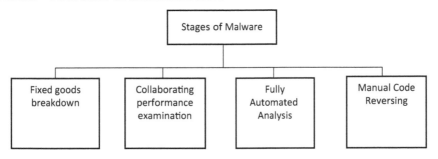

- *Fixed Goods Breakdown:*
  It incorporates strings inserted within the malware code, header elements, and hashes, metadata, etc.
- *Collaborating Performance Examination:*
  The behavioral examination is utilized with malware tests running in a lab. Investigators look to get the sample's registry and record framework, then prepare and arrange exercises. They may conduct Memory Forensics to understand how the malware employs memory. If the examiners suspect the malware has specific capabilities, they can set up a recreation to test their hypothesis.
- *Fully Automated Analysis:*
  An introductory and speedy survey of suspicious records is a completely automated examination and is the best way to handle malware at a scale.
- *Manual Code Reversing:*
  This analysis uses reverse-engineered code generated from debuggers, disassemblers, decompilers, and specialized instruments to interpret scrambled information, decide the rationale behind the malware calculation and reveal any hidden capabilities that the malware has not yet shown.

10.2.3 MALWARE ATTACKS

This is when cyber criminals activate destructive code found in someone's device. The user is then deprived of their data and must let specific data corrupt or pay a ransom.

10.3 MALWARE FORENSICS

Malware is a suspicious computer program that assaults a framework program, where forensics reconnoiters a noxious assault. Forensics is done to know the effect of the specific malware. The Microsoft Windows [3] framework is the most broadly used OS, making it the target of malware developers. This shows up in multiple executables such as BAT scripts, VB Scripts, Java Scripts, and more.

10.3.1 ADVANCED MALWARE

Advanced malware [4] alludes to Progressed Tireless Risk (PTR). Malware strains are built with advanced capabilities for disease, communication, control, development, information infiltration, or payload execution. Live forensics is utilized to assemble framework data from the contaminated framework. Typically, this is the reason why protecting unstable information is vital for malware examination. On the off chance that a smash dump is not performed, to begin with, at that point, the framework state might alter.

10.3.2 MEMORY FORENSICS

The method of acquiring and examining the critical memory of the device is Memory Forensics. In the live response process, investigating an intrusion or a malware infection is critical. It makes it possible to gather and analyze volatile objects that only exist in memory in certain instances.

Prominence of Memory Forensics
- It contains the complete data which caused the incident.
- Through the utilization of Memory Forensics, an advanced examination can notice protected memory.

10.3.3 CASE STUDY 1: RATIONALIZATION THE ASSORTMENT AND RESULTS [5]

A test center was organized:

1. Virtual environment utilizing VMware Workstation.
2. Isolated organization; utilizing a fake network to recreate a live organization.
3. Windows 7 Professional 32-bit working framework.
4. 4GB RAM.
5. Test Output.

Process:
   The .bat file is run, wait for the execution to stop, collect the result.

Scrutinize the Result:
   Amount produced from the query.

```
alosh66.no-ip.info
------------------------------------------
Record Name . . . . . : alosh66.no-ip.info
Record Type . . . . . : 1
Time To Live  . . . . : 85041
Data Length . . . . . : 4
Section . . . . . . . : Answer
A (Host) Record . . . : 127.0.0.1
```

Fig. output

The example Figures 10.1, 10.2, and 10.3 show mining of executable records out of RAM dump using the tool.

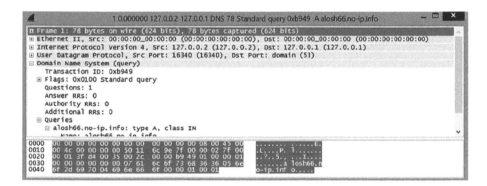

FIGURE 10.1   File extractor.

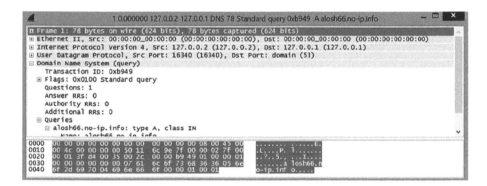

FIGURE 10.2   Input from the scan.

```
█ Administrator Command Prompt

F:\collection\analyze>volatility-2.3.1.standalone.exe -f 20140607.mem --profile=Win7SP0x86 procmemdump -p 4040 -D ./
Volatility Foundation Volatility Framework 2.3.1
Process(V) ImageBase  Name               Result
━━━━━━━━━━━━━━━━━━━━━━━━━━━━━━━━━━━━━━━━━━━━━━━━━━━━━
0x8441c030 0x001c0000 IEShims.exe        OK: executable.4040.exe

F:\collection\analyze>_
```

FIGURE 10.3 Final output.

Decision
Examining the initial data begins with assembling documents to evaluate. The programming toolkit provides an easy way to collect data for specialists.

10.3.4 CASE STUDY 2: FRAUDSTER TRIES TO ACCESS CLIENT'S SUPER FUNDS AFTER EMAIL HACKED

A man who works overseas used to receive instructions from a specific client through email. The client's email had been hacked and sent emails to an end party. The employee finds the hoaxer. Upon this the end party requested the email to be changed by a small letter in that. After the attack, the man receives an email asking about his bank balance. The hoaxer sent an email demanding $500,000 to be deposited. This continues, and man could not access them directly. So, they stopped all the issues.

Preventing This Type of Fraud
   • This is one of the most common ways fraudsters get money. They hack into people's email accounts and use them to send fraudulent emails.
   • Always be suspicious if you get an email asking for a considerable sum to be paid to a modern or foreign bank account – particularly on the off chance that you have been asked to alter the client's contact details by a single digit or character. This ploy is used to avoid you making contact by phone, email, or text message to the client.
   • A short call to your client, using the contact details you have on record, confirms if you are dealing with a fraudster.

10.4 MALWARE FORENSICS TOOLS

Malware Tools are generally two types.

   1. Static Analysis Tool
   2. Dynamic Analysis Tool

10.4.1 STATIC OR BASIC ANALYSIS TOOLS

1. Tool 1

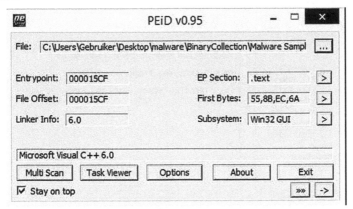

This tool is used to extricate regular packers [6]. The current interpretation of a portable executor identity can diagnose 500 distinctive marks in a portable executor [7] document.

2. Tool 2

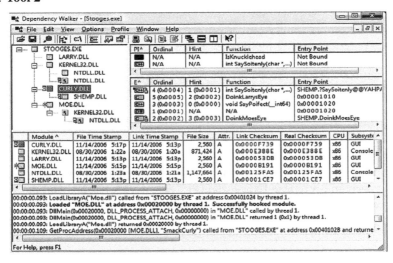

It is used [8] to list all the functions of a module. It displays complete information about those files together with the file path, etc.

3. Tool 3

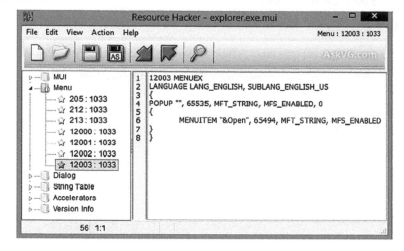

Sometimes called Supply Hackers [8], it is also an open-source piece of software used to extract all resources.

4. Tool 4

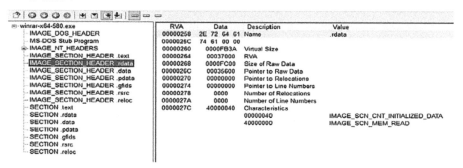

Portable Originator View [9] is a free and simple application to peruse the data put away in a Portable Executable (PE) document's headers and the various areas of the record.

10.4.2   DYNAMIC ANALYSIS TOOLS

1. **Wiresharks**

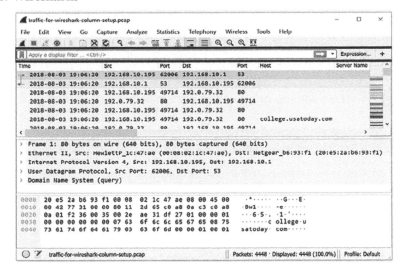

Wiresharks [10] is used to examine what is presently happening and catch packets to documents. This can be utilized for live bundle catching.

2. Internet Sim Tool

Internet Sim Tool [11] is a Linux-based tool used to simulate common internet services. It also answers service requests.

3. Net_Cat

```
root@kali:~/Documents/PWK/lab/10.11.1.13# nc -lvnp 31337
Ncat: Version 7.70 ( https://nmap.org/ncat )
Ncat: Listening on :::31337
Ncat: Listening on 0.0.0.0:31337
Ncat: Connection from 10.11.1.13.
Ncat: Connection from 10.11.1.13:3196.

root@kali:~/Documents/PWK/lab/10.11.1.13# netcat -lvnp 31337
Ncat: Version 7.70 ( https://nmap.org/ncat )
Ncat: Listening on :::31337
Ncat: Listening on 0.0.0.0:31337
Ncat: Connection from 10.11.1.13.
Ncat: Connection from 10.11.1.13:3198.
```

This is used in TCP and UDP connections and performs multidisciplinary operations [11].

4. RegShot

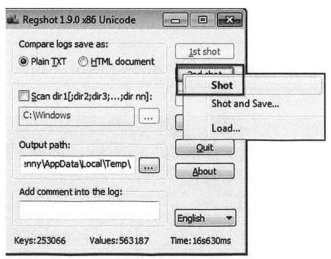

It is used for previewing records which can then be compared to the current state of the registry [11]. Similarly, it can view the changes made to the registry.

10.5   PROCEDURE MONITOR

Procedure monitor, or Process Monitor, tool and is used to monitor the whole windows. The tool is some inheritance riggings from File Monitor and Register Monitor. Added features for straining of facts and start time cataloging.

10.5.1   OPEN-SOURCE MALWARE FORENSICS TOOLS

Examiners utilize free examination devices to secure from and anticipate future assaults and to share information. By utilizing free malware tools [12], investigators can test, characterize, and report distinctive variations of malevolent acts while learning about the assault's lifecycle.

1. Cuckoo Sandbox

It provides suitable comments regarding file behavior in out-of-the-way environments.

2. Yara Rules

```
rule rogues
{
    meta:
        created  = "09/11/2017 00:00:00"
        modified = "09/11/2017 17:12:07"
        author   = "Metallica"
        Vendor   = "PC Smart Cleanup"
    strings:
        $textstring1 = "smart" ascii wide nocase
        $textstring2 = "cleanup" ascii wide nocase
        $textstring3 = "longrun" ascii wide nocase
    condition:
        $textstring1 and @textstring2 and $textstring3
}
```

Yara, which stands for "Yet Another Recursive Acronym," allows researchers to distinguish and sort ostensibly comparable deviations of malware and can be unified to use within Cuckoo.

10.5.1.1 Example of Advanced Malware (APT)

APT examples include many attacks with a real-world scenario, in which it describes the most common are phishing and fraudulent data. The stages include finding the target, organization, and required tools, then testing for discovery and deploying the analyzed intrusion. Expand the access and achieve authorizations.

If your organization sees an irregular expansion of login attempts, it may be another clue that an APT is attempting to invade your organization. These signs could be people working late to meet deadlines or a lesser attack attempting to force its way in. Regardless, organizations should be watchful, particularly if late-night logins are joined with a portion of different pointers.

10.6 CONCLUSIONS

If a system is affected by malware, then a malware forensic examination can help find it. Following such methods, the malware can be traced while documenting most trace evidence related to a malware incident. Then repeated forensic examinations are done to recover them.

REFERENCES

1. "Crowdstrike," accessed August 26, 2021, http://www.crowdstrike.com/.
2. "Introduction to Malware and Malware Analysis," Quickheal R & D Lab.
3. Cameron, H., et al. *Malware Forensics Field Guide for Windows Systems*, Elsevier, 2012.
4. Kunwar, R. S., Sharma, P. "Malware Analysis: Tools and Techniques," *Second International Conference*, 2016.
5. Joyce A. V. "A Technique for Streamlining Data Collection," *42nd Annual Communicating Nursing Research Conference Western Institute of Nursing*, 2009.

6. "PEiD," accessed August 26, 2021, http://www.softpedia.com/get/Programming/Packers-Crypters-Protectors/PEiD-updated.shtm/.
7. "Dependency Walker 2.2," accessed August 26, 2021, http://www.dependencywalker.com/.
8. "Resource Hacker™," accessed August 26, 2021, http://www.angusj.com/resource hacker/.
9. "Five PE Analysis Tools Worth Looking At," accessed August 26, 2021, https://blog.malwarebytes.com/threat-analysis/2014/05/five-pe-analysis-tools/.
10. "Wireshark," accessed August 26, 2021, http://www.wireshark.org/.
11. "INetSim: Internet Services Simulation Suite," accessed August 26, 2021, http://www.inetsim.org/.
12. Talukder, S., et al. "A Survey on Malware Detection and analysis Tool," *International Journal of Security and Its Application*, 10.

11 Network Intrusion Detection System Using Naïve Bayes Classification Technique for Anomaly Detection

Sam Goundar
RMIT University, Vietnam

Manveer Singh and Rahul Chand
The University of South Pacific, Fiji

Akashdeep Bhardwaj
University of Petroleum and Energy Studies,
Dehradun, India

CONTENTS

DOI: 10.1201/9781003140023-11

11.1 INTRODUCTION

The dramatic growth in computer resources has seen a variety of network-based applications being developed to provide services in many different areas, for example, we have seen the impressive impact of e-commerce technology in our fast-moving society, an article from the daily newspaper also highlighted that the introduction of public web services provided by the Government is also a step up for our economy in the past recent months (Bolatiki, 2018), along with other commercial and non-commercial based services concerning computer usage. In today's globalized, fast-changing business environment, staying secure is not just the sole responsibility of the IT department but for the whole organization. It is critical for the survival of any organization or public entity. With the tremendous growth of network-based services and sensitive information on networks, network security is getting more important than ever. Intrusion poses a serious security risk in a network environment. The ever-growing new intrusion types present a serious problem for their detection. An Intrusion Detection System (IDS) inspects the activities in a system for suspicious behavior or patterns that may indicate system attack or misuse. Network Intrusion Detection Systems (NIDS) are essential tools for network system administrators to detect various security breaches inside an organization's network (Wahl, 2016). A NIDS monitors and analyzes the network trace entering into or exiting from the network devices of an organization and raises alarms if an intrusion is observed. As mentioned in the Abstract above, this chapter is focused on the use of anomaly detection systems through Data Mining techniques.

Most traditional NIDSs detection systems attempt to match patterns and signatures of already known attacks in the network traffic, whereby a constantly updated database is usually used to store the signatures of known attacks (Rahul, Vinayakumar, Soman, & Poornachandran, 2018). It cannot detect new attacks until it is trained for them. However, an anomaly detection system attempts to identify behavior that does not conform to normal behavior through the use of the Naïve Bayes Classification technique for network intrusion detection (Gujar & Patil, 2014). In this research paper, we apply one of the efficient data mining algorithms called Naïve Bayes Classification for anomaly-based network intrusion detection. This research will help understand the importance of IDS in organizations, it will help illustrate and understand the connection between data mining and IDS in detecting anomalies. Over the past few decades, data mining procedures have gained a lot of popularity around the intrusion detection field in which researchers try to address the shortcomings of knowledge base detection systems. This has prompted the requisition of different supervised and unsupervised learning with the end goal of intrusion detection. Although there has been a lot of research done in this field, there still exist problems in identifying intrusions in the incoming traffic of the network. These can be in the form of lack of accuracy of detection rate, or it can behave differently for different types of attacks which lead to work well for one type of data but not for others, hence, result in unbalanced detection rates and high false positives. Apart from that, data sets have redundant input attributes as well as examples in the training data, which leads to situations in which IDS cannot detect anomalies as it only looks at real-time data which are present in the network and are not encrypted. There are no illustrations of hidden patterns to find abnormalities in the network which the system administrators can pick out because most of the data subside within the system and system administrators are unaware of any real threat or vulnerability in their network.

This research will also allow us to understand Naïve Bayes Classification and ideas generated in network security using data mining. It will also help identify anomalies and hidden patterns using data mining using IDS. The motivation to conduct this research and write the research paper was because of the continuous data breaches that occur in large corporate firms, even with IDS in place. This is because the IDS is in place is not monitoring real-time data and is not looking at any sort of anomalies or hidden patterns, and even the encrypted data flowing through the network is not analyzed in most firms. The problem here is that intruders can easily gain access to the network without being detected and communicate through encryption. When this happens, it will not be picked up by the IDS as it is not made to look into anomalies and hidden patterns flowing through the encrypted data in the network.

The overall research objectives of this chapter are as follows:

- To understand the importance of IDS in organizations.
- To understand the connection between data mining and IDS in detecting anomalies.
- To understand Naïve Bayes Classification and applying ideas which integrate with Network Intrusion Detection System.
- To identify anomalies and hidden patterns using Data Mining techniques in Network Intrusion Detection System.

And we attempt to address the following research questions:

- How can real-time data be gathered and prepared from the network, which will be used in data mining to analyze the network for intrusions?
- How can Naïve Bayes Classification help in finding anomalies and patterns in the system to create a trustworthy investigation of a network to prevent intrusions?
- What are the ways in which Data Mining techniques can effectively prevent attacks using IDS in a network?
- How can we effectively test and implement an IDS using a Naïve Bayes Classification technique?

Our research is based on the works of Mukherjee and Sharma (2012) and then extended and customized to fit the requirements of our network security investigation. The research paper published by Mukherjee and Sharma (2012) discusses the use of data mining which uses the Naïve Bayes Classification technique with Feature Reduction to identify anomalies through prioritization of IDS features. Symantec Corporation, in a recent report, uncovered that the number of phishing attacks targeted at stealing confidential information such as credit card numbers, passwords, and other financial information are on the rise, going from 70 million attacks in

June 2018 to over 150 million in less than a year (Translated by ContentEngine, 2019). It is therefore important that NIDSs are accurate in identifying attacks and quick to train and generate as few false positives as possible. This chapter starts off with a comparative study of anomaly detection schemes for identifying novel network intrusion detections. Experimental results have been gathered and analyzed, and the Naïve Bayes classifier model is seen to surpass detections of unusual activities and network intrusions. Section two describes IDS in general and relates our research to other works done in the same field. Section three presents an overview of the methods used in this research and also the model used for insight which presents the Modeling and Theory used for research compilation. Section four and five talk about the frequently occurring network attacks along with the evaluation and discussion of each of the results in section five. Finally, section six provides the concluding remarks and future scope of this research work.

11.2   LITERATURE REVIEW

Intrusion is a type of attack on a computer or a network-based environment that attempts to bypass the security countermeasures of a system and gain access to the platform. An IDS inspects the activities in a system for suspicious behavior or patterns that may indicate system attack or misuse. It has been known that most NIDS detection attempts to match patterns and signatures of already known attacks in the network traffic. A constantly updated database is usually used to store the signatures of known attacks (Bace, 2000). It cannot detect new attacks until trained for them.

Anomaly detection attempts to identify behavior that does not conform to normal behavior. This technique is based on the detection of traffic anomalies. The anomaly detection systems are adaptive in nature, they can deal with a new attack, but they cannot identify the specific type of attack (Markou & Singh, 2003). In this chapter, we review the performance of classifiers when trained to identify signatures of specific attacks. These attacks are discussed in more detail in the following section of the literature review.

One of the earliest works found in literature used Artificial Neural Networks (ANN) with enhanced resilient backpropagation for the design of such an IDS. This work used only the training dataset for training (70%), validation (15%), and testing (15%).

The increasing hacking incidents made IDS an essential objective in any network. Out of many algorithms, the Naïve Bayes is one of the best classification models that predicts intrusion very fast since the algorithm is less complicated (Bace, 2000). A behavior model is introduced that uses Bayesian techniques to obtain model parameters with maximal a-posteriori probabilities (Mukherjee & Sharma, 2012).

Another area of concern within intrusion detection is to be able to identify a specific intrusion. For example, has there been a misuse intrusion (misuse detection) and/or anomaly intrusion (anomaly detection)? According to Sisodia, Sharma, and Pandey (2012), misuse detection attempts to match patterns and signatures of already known attacks in the network traffic, while anomaly detection attempts to identify behavior that does not conform to normal behavior.

11.2.1 Naïve Bayes Classification Data Mining Technique

There are many Naïve Bayes Classification Data Mining Techniques. We list below some commonly used techniques:

Audit Data Analysis and Mining (ADAM) is an intrusion detector built to detect intrusions using data mining techniques. It first absorbs training data known to be free of attacks. Next, it uses an algorithm to group attacks, unknown behavior, and false alarms (Basak, 2014).

Intrusion Detection using Data Mining Technique (IDDM) is a real-time NIDS for misuse and anomaly detection. It applies association rules, meta-rules, and characteristic rules. It employs data mining to produce a description of network data and uses this information for deviation analysis (Sisodia et al., 2012).

Mining Audit Data for Automated Models for Intrusion Detection (MADAM ID) is one of the best-known data mining projects in intrusion detection. It is an off-line IDS to produce anomaly and misuse intrusion detection models (Lee & Stolfo, 2000).

The authors (Zhang, Liu, Jia, Ren, & Zhao, 2018) applied backpropagation artificial neural network models into intrusion detection, which makes IDS more efficiently adapt to new environments and respond to new types of attacks. Due to the large size of the network dataset, manual tagging would consume a lot of time and effort; thus, clustering methods are introduced into the dataset classification.

According to Zhang et al. (2018), although the data mining method has good adaptability to new attack types, it is often higher in time consumption. Principal Component Analysis (PCA) is a commonly used dimensionality reduction technique. It uses an orthogonal transformation to convert a set of related variables into a set of linearly uncorrelated variables, where the first principal component has the largest variance.

The Bayesian method has been incorporated with data mining techniques to a certain extent and is faster than other classifiers because it is a classifier based on conditional probability (Lee and Stolfo, 2000). The researcher (Panda, 2010) used a Bayesian belief network with a local genetic search for intrusion detection. He concluded that his model is able to detect new attacks as well as experienced attacks.

Barot and Toshniwal (2012) proposed a new hybrid model using Naïve Bayes Probability Theorem and Decision Table, whereby Naïve Bayes theorem gives a probability which is subsequently classified into a decision table from which the nature of the threat is derived. Farid (2010) described an algorithm for detecting adaptive network intrusions using Bayesian classification and decision trees to improve accuracy. This method eliminates false-positive levels of various types of network intrusions at an acceptable level. It also removes all redundant attributes from the learning dataset as well as inconsistent attributes.

An adaptive Network Intrusion Detection System was proposed by Karthick, Hattiwale, and Ravindran (2012). This system followed a two-stage approach to architecture. Where a probabilistic classifier was used in the first stage to predict traffic intrusions, and in the second stage, a traffic model based on Hidden Markov Model (HMM) was used to eliminate potential attacks associated with IP addresses. They state that "any activity aimed at disrupting a service or making a resource unavailable or gaining unauthorized access can be termed as an intrusion." Examples include buffer overflow attacks, flooding attacks, system break-ins, etc.

Researchers (Tsuruoka & Tsujii 2003), in their papers, attempt to exploit unlabeled data. They use the Naïve Bayes classifier and combine it with the well-established Expected Maximization (EM) algorithm. In the iteration cycle of the EM algorithm, a category distribution limit is implemented.

Data Mining is also used in the medical field, an article by Subbalakshmi, Ramesh, and Rao (2011) discusses how researchers developed the Heart Disease Prediction and Decision Support System using the Naïve Bayesian Classification technique. The program collects hidden knowledge from a list of chronic heart diseases. This model would respond to complex questions, each with its own strength in terms of ease of model analysis, access to detailed data, and reliability.

In an article on network security, researchers (Chiche & Meshesha, 2017) suggested a NIDS system based on the algorithm of Naïve Bayes Classification. They built their own database by collecting real-time data. The results show that the proposed approach is best suited to real-time data. This work can be expanded by using various classifiers in real-time environments and can also extend this approach to detect different types of networking attacks in real-time.

Using another method, researchers (Govindarajan & Chandrasekaran, 2011) use more than one classifier for the training phase, i.e., a collection of classifiers. The use of classifier ensembles has proven to be an effective way to improve IDS. For example, the researchers used a new hybrid Radial Base (RB) and Support Vector Machines (SVM), which uplifts overall categorization capability. It also saves storage space since the discretized data requires less space.

In another research paper, researchers (Panda & Patra, 2009) introduced a method to apply machine learning algorithms to establish patterns for the detection of network intrusion. The machine collects the packets from the network traffic during the initial phase. The features for each link are built from the captured network traffic by the pre-processors. Then training data is fed in pattern building unit, which then constructs patterns. Instead of using the patterns developed in the initial phase, the connections are marked as specific intrusions or regular traffic.

The use of cell data by IDS to improve detection was put forward (Warrender, Forrest, & Pearlmutter, 1999). They state that "several intrusion detection methods are based upon system call trace data." In their paper, they discuss how IDS uses system cell data that is generated by different programs which compare and analyze simple enumeration of observed sequence.

An article published under the security and communications network field examines IDS by sliding windows to evaluate a regular sequence database in order to create a database for checking against sample instances (Malik, Shahzad, & Khan, 2015). They then used a similar method to compare windows against the database in the sample instances and identify instances in the usual sequence database according to those. For each call made by a process, the function requires sequential analysis of a window of system calls. It requires a large list of standard system call trace sequences to be preserved.

According to Sethi, Mishra, Solanki, and Mishra (2009), many intrusions occur via computer networks to attack their targets using network protocols. Sethi et al. (2009) proposed a new model for the creation of an intrusion detection method in immunology and called it "Danger Theory." This theory suggests that the immune system responds to threats based on the connection between specific (danger) signals and offers alert and mitigation suggestions. Determining the difference between normal and potentially harmful behavior is the central problem with computer security.

Association Rule Mining is one of the most common, widely used, and fundamental techniques in data mining, which discovers the connection between items in a large database. Besides this, it is designed specifically for use in data analysis. The rule of association is available in either the Apriori model or the model of frequent growth pattern (Arvidson & Carlbark, 2003). The association rule format is "X=>Y," where X and Y are the item sets that can be translated as X itemset existence indicates the presence of Y itemset in a particular transaction.

In one of the papers published under the field of data mining, the authors proposed a clustering framework for intrusion detection for the identification of anomalous events based on simple KMeans for the analysis of network flow information using any flow data attribute such as IP address, path, anomaly detection protocols (Chary, 2012). A NetFlow collector to collect network flow information from network traffic is utilized in this method. This set of data is pre-processed and indexed on the basis of key parameters such as IP addresses, ports, and protocols.

Information mining-based IDS models provide primary data collection, data mining, pattern matching, and decision-making (Salo, Injadat, Nassif, Shami, & Essex, 2018). According to the authors, "the mining engine is converting network data into data files." First, data records are analyzed by the mining algorithms to learn ordinary and abnormal record laws. If there are new rules, the rules are generated and stored. In the next step, compare the rules with newly generated rules, then evaluate the rules through the intelligence decision module to determine whether or not an intrusion has occurred.

Deng, Liu, and Zhou (2015) presented the Genetic Algorithm (GA) along with the initial population and selection modifications. The GA is used in audit files to automate the quest for attack scenarios. This includes the sub-set of potential threats available in a reasonable preparation time in the evaluation report. They used the data set (NSL-KDD99) of the network security lab knowledge discovery and data mining. By combining the IDS with the GA, the network intrusion detection model's detection rate increases performance, and the false-positive rate decreases resulting in a better IDS.

11.2.2 Networking Attacks

Each attack type falls into one of the four main categories according to the motives and mediums of attacks, according to Sisodia et al. (2012). The four categories are:

1. Denials-of Service (DoS) attacks have the goals of limiting, overloading or denying services provided to the user, computer, or network. A common tactic is to severely overload the targeted system. (e.g., Apache, Smurf, Neptune, Ping of death, back, mailbomb, udpstorm, SYNflood, etc.).
2. Probing or Surveillance attacks have the goal of gaining knowledge of the existence or configuration of a computer system or network. Port Scans or

sweeping of a given IP address range typically fall in this category. (e.g., saint, portsweep, mscan, nmap, etc.).

3. User-to-Root (U2R) attacks having the goals of gaining root or super-user access on a particular computer or system on which the attacker previously had user-level access. These are attempts by a non-privileged user to gain administrative privileges (e.g., Perl, xterm, etc.).

4. Remote-to-Local (R2L) is an attack in which a user sends packets to a machine over the internet, which the user does not have access to in order to expose the machine vulnerabilities and exploit privileges which a local user would have on the computer (e.g., xclock, dictionary, guest password, phf, sendmail, xsnoop, etc.).

11.3   RESEARCH METHODOLOGY

Naïve Bayes Classification algorithm is a probabilistic classifier/class of algorithms that bases probability models that incorporate assumptions and probabilities. This technique constructs models that assign class labels to problems instances. Here, it assumes that the value of a particular feature is independent of the value of any other feature, meaning no one entity entitles the value of another, independent, given the class variable (Nielsen & Jensen, 2013). For example, a shoe is wearable if it is in good condition, it has laces on it, has a sole, and your size. A Bayes classifier considers each of the above attributes (features) to assume that the shoe is, in fact, wearable independently regardless of the correlations between condition, lace, sole, and size features. In Bayesian classification, a screening of the dataset is required, and if there are new elements introduced, then another screening is required. Then we hypothesize that the given data belongs to a particular class. We then calculate the probability for the hypothesis to be true. This is among the most practical approaches for certain types of problems. Thus, a Bayesian network is used to model a domain containing uncertainty (Chai, Lei, & Fang, 2017).

11.3.1   INTRUSION DETECTION METHODOLOGIES

For this chapter, the information gathering process and investigations were surrounded by the quantitative paradigm, and the methodologies based on our research are numerical and are used for experimental purposes. For the information gathering process, network-based IDS methodology from online sources was experimented with, which is either signature-based or heuristic. The main focus area of this was to build an enhanced understanding of the IDS being used by large corporate organizations and taking out the flows of the system. According to Rao and Nayak (2014), a Network IDS (NIDS) is a stand-alone mechanism to screen movement all around that system (Figure 11.1).

Network traffic is analyzed against a dataset, and patterns are built from which the theorem is able to classify, detect which patterns are legit and which are not, which in turn shoots an alert. The results of the Naïve Bayes classifier are often correct. The work reported in Shyara Taruna and Hiranwal (2013) examines the circumstances under which the Naïve Bayes classifier performs well and why. It states that the error

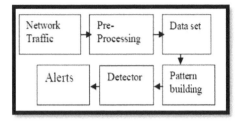

FIGURE 11.1 Framework of intrusion detection model (IDM).

$$p(A|B) = \frac{p(B|A) - P(A)}{P} \quad (B)$$

FIGURE 11.2 Formula: Bayes Theorem to calculate conditional probability.

is a result of three factors: training data noise, bias, and variance. Training data noise can only be minimized by having good training data. Training data must be divided into various groups. Bias is the error due to groupings in the training data being very large. Variance is the error due to those groupings being too small (Figure 11.2).

Where:

- P (H) is the probability of hypothesis H being true. This is known as the prior probability.
- P (E) is the probability of the evidence (regardless of the hypothesis).
- P (E|H) is the probability of the evidence given that the hypothesis is true.
- P (H|E) is the probability of the hypothesis, given that the evidence is there.

11.3.2 KNOWLEDGE DISCOVERY IN DATABASES (KDD) CUP 1999 DATASET METHODOLOGIES

In our research, the main challenge was to find a large enough organization which allowed us to evaluate their network data. This was difficult as most organizations do not even allow access to external users of any sort as it becomes a data breach. Due to that reason, one of the best choices which we had was to opt for online experimental data to allow for this research to take place. KDD Cup 1999 dataset was used for training and testing in our NIDS for this research. All the attributes of the dataset, conversion, and manipulation techniques applied to the dataset are discussed later.

In academic research, it is stated that anomaly detection is more effective due to its hypothetical advantage for tracing new attacks. Through the use of a classifier, which is fit for recognizing common attacks, the 1998 DARPA Intrusion Detection Evaluation Program was started and overseen by MIT Lincoln Labs (Abraham, Grosan, & Chen, 2005). This was done through various verification processes, which included the test data to be reviewed and evaluated in intrusion detection. This brings a standard set of information to be examined, which is challenged by the 1999 KDD

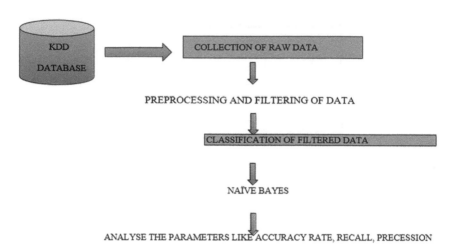

FIGURE 11.3 The network intrusion steps undertaken.

intrusion detection challenge. For each of the connections, various quantitative and qualitative features are obtained with normal and attack data.

One of the major points of interest for our research to use this dataset is that it is equipped with huge data mining capabilities and is also used in other university courses offered in data mining. As mentioned earlier, the essential goal of this research work is to focus on anomalies detection in the interpolation of the network using the KDD'99 datasets to evaluate identifications. To achieve this, quantitative dependency of the methodology upon the 5% of data from the total KDD'99 is used due to the scope and timeliness of the research. Considering, ascertained entropies will be near 0, and the data increase will be near 1 (Figure 11.3).

We collected the dataset from an online source since it was difficult to find an organization which will allow just in-depth scanning of their network data. However, only 5% of the total dataset was used. The dataset selected was the KDD Cup 1999, which is a standard dataset used in data mining for academic research.

11.3.3   ATTRIBUTES OF KDD' CUP 1999 DATASET

A total of forty-one (41) features are selected from the KDD' Cup 1999 dataset focusing only on 5% of the total dataset. Among these features, a total of nine (9) focuses on the basic features of a network packet. A total of twelve (12) are based on content features, and eight are focused on traffic features, and nine are the host primarily based features. To observe the different attacks based on the features selected, the dataset is divided into two classes.

1. Same Host: It is determined through the research of past details, which handles the flow association and administration of the dataset.
2. Same Service: Focuses mainly on the connections of the past with the current connections.

TABLE 11.1
List of Symbolic Features with Details on Network

Number	Feature	Attributes	Details
1	Type	4	Is it TCP, UDP, ICMP?
2	Service	16	Services available to be utilized
3	Flag	6	Status of connection?
4	Land	1	Looks at the destination and host connection
5	Logged In	1	Number of attempts needed
6	In_hot_login	1	Is it from a HOT list?
7	In_guest_login	1	Is it from a GUEST list?

Both draw a picture of how slow attacks that threaten the hosts or ports amount to larger resources needed.

11.3.4 SYMBOLIC FEATURES OF KDD' CUP 1999 DATASET

A total of seven symbolic features in which four are binary, and the other contains more than two attributes. The latter will be converted to numerical features in our research paper. The following are the symbolic features listed in Table 11.1.

11.3.5 NUMERICAL FEATURES OF KDD' CUP 1999 DATASET

Our dataset includes a total number of 32 features with different ranges, which are described in Table 11.2.

TABLE 11.2
List of Numerical Features with Details on Network

Number	Features	Details
1	Duration	Information of the time period is listed here
2	Src_bytes	Information about the amount of data in bytes
3	Dst_bytes	About the amount of data sent from destination
4	Wrong_fragment	The number of wrong fragments in the network
5	Urgent	Number of packets which are urgent
6	hot	Number of Hot indicators in the network
7	Num_failed_logins	Number of failed attempts in logging in
8	Num_compromised	Conditions in a network which are compromised
9	Root_shell	0 till 9 and reflects if the root_shell is logged
10	Su_attempted	0 or 1, command if su_attampted been activated
11	Num_root	The access level for root in the network
12	Num_file_creations	Gives number of files in the network to be created
13	Num_shells	Looks at the prompts in the shell
14	Num_access_files	Operations performed under access control files
15	Num_outbound_cmd	Looks at the outbound files

(Continued)

TABLE 11.2   (Continued)

Number	Features	Details
16	Count	Number of connections in the network
17	Srv_count	Count based on service
18	serror_Error	Looks at the percentage of SYN errors
19	Srv_serror_error	Looks at the percentage of SYN errors
20	Rerror_rate	Looks at the percentage of REJ errors
21	Srv_rerror_rate	Looks at the percentage of REJ errors
22	Same_srv_rate	Percentage that are dedicated to the same service
23	Diff_srv_rate	Percentage dedicated to the different services
24	Diff_srv_host_rate	Percentage that are dedicated to the host service
25	Dst_host_count	The number of hosts in the current destination
26	Dst_host_srv_count	The number of services in the current destination
27	Dst_host_same_srv_rate	Based on the same service
28	Dst_host_diff_srv_rate	Based on the different service
29	dst_host_same_src_port_rate	Same source ports of the connection
30	dst_host_srv_diff_host_rate	Different source ports of the connection
31	dst_host_serror_rate	Network connection error SYN in destination port
32	dst_host_srv_serror_rate	Network connection error REJ in destination port

11.4   RESULTS

In this section, we will look at the procedures of how the Network Intrusion Detection System utilizes the KDD dataset. For this large amount of dataset, we had to use the NetBeans application for data mining of the Naïve Bayes Classification algorithm. The experiment was only focused on 5% of the KDD Cup 1999 data set. In this research, we evaluated the performance of the entire network system by false-positive rate. To define false-positive rate in network terms, Goeschel (2016) states that "false-positive rate is the number of normal connections that are misclassified as attacks divided by the number of normal connections in the dataset" (Goeschel, 2016) (Figure 11.4).

As mentioned, our research scope focuses on the false-positive, which analyzes the network and indicates that something is not right when it is right, for example, highlighting a price of code as malicious code. However, the code is genuine.

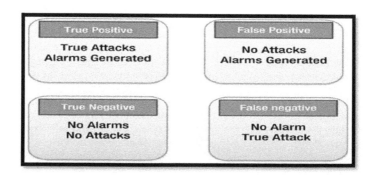

FIGURE 11.4   Different IDS/alerts.

FIGURE 11.5 The main file used. (KDD Cup'99).

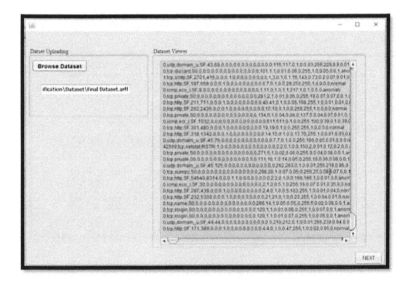

FIGURE 11.6 The content of the dataset.

The following figures show the demo undertaken to find the outcome (test data) using NetBeans (Figure 11.5).

Once we click on the main file and click the run file button. We will need to browse the dataset (Figure 11.6).

We have used the KDD dataset, in this screen, we will be able to see the attributes of the dataset (Figure 11.7).

We will be able to see the pre-processing of the data set. Now we are ready for the IDS implementation using Naïve Bayes Classification algorithm (Figure 11.8).

FIGURE 11.7   The filtration/pre-processing of the dataset.

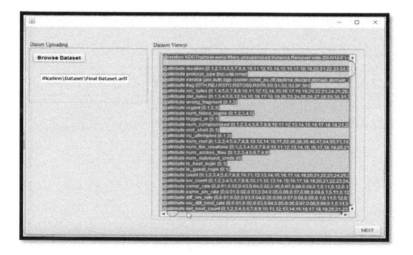

FIGURE 11.8   Attributes of KDD dataset.

Once you can see the content of the dataset, the class attribute (highlighted) of the dataset is normal traffic or anomaly traffic. It will classify the data into two categories that are normal or anomaly (Figure 11.9).

This will basically replace the missing values in the dataset and do the conversion of numeric data into nominal data (Figure 11.10).

Now, this dataset will be used in the classification of Naïve Bayes (Figure 11.11).

As Naïve Bayes algorithm is based upon the conditional probability. That is that it uses the Naïve Bayes theorem for classification purposes. In this case, we have a total number of 213 instances, out of which Naïve Bayes has correctly classified 201 instances (94% of accuracy), and 12 are incorrectly classified (6% of non-accuracy).

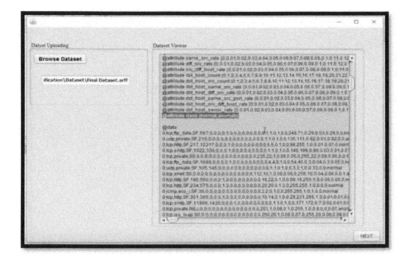

FIGURE 11.9   Shows the content of the dataset.

FIGURE 11.10   Shows the filtration/processing of the dataset.

We can also use the complexity parameters and check the error rate for the evaluation of the Naïve Bayes model. At the bottom, you can see the class detailed parameters, which are defined using the confusion matrix. In our case, some of the diagonal elements of the confusion matrix are 119 added with 82, which equals 201, which means that correctly classified instances are equal to the sum of the diagonal elements of the confusion matrix. Whereas non-diagonal elements sum is equal to incorrectly classified instances. We can see the evaluation of the false-positive rate, which you can see in the "FP Range" of the screen of the NetBeans application, which is 0.071 (Figure 11.12).

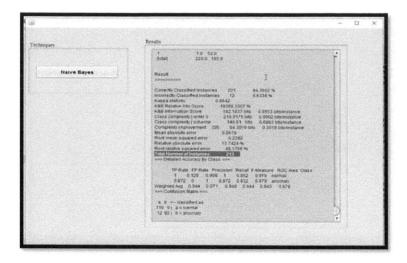

FIGURE 11.11   The classification results of the KDD dataset.

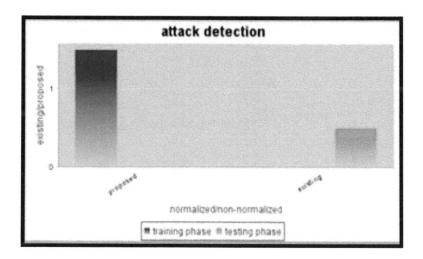

FIGURE 11.12   The attack detections in graphical format using NetBeans.

In the attack detection graph, we can clearly see evaluations of the proposed (Blue) and the existing (Green) highlighting the attack detection. Non-normalized data is depicted in the X-Axis, whereas the existing proposed data is depicted in the Y-Axis.

11.4.1 ANALYSIS

- How can real-time data be gathered and prepared from the network, which will be used in data mining to analyze the network for intrusions?

- Server logs can be fed into a data warehouse in real-time for gathering information about a network's traffic, to and for which can be sampled to see what legitimate and illegitimate network accesses are. Real-time data is used in this instance is essential as data and users move in real-time hence acquiring details there and then is essential on determining what is what and who is who. This is essential in classifying a matrix of authenticated and unauthentic users and locations which is later used as a reference for cross-matching/referencing.
- How can Naïve Bayes Classification help in finding anomalies and patterns in the system to create a trustworthy investigation of a network to prevent intrusions?
- Naïve Bayes algorithm classifies pings and datasets into segments and groups them as either trustworthy or anomalies. For example, anomalous traffic in a computer network could signal the failure of one or more devices or that a hacker has penetrated the network. Naïve Bayes will not classify this as an attack rather an anomaly which the user has to inspect and deduce results off. Naïve Bayes acts as an early warning system that cross-references given datasets against collected datasets to distinguish what is what and so forth.
- What are the ways in which Data Mining techniques can effectively prevent attacks using IDS in a network?
- Data mining is a step in the process of KDD involving the application of data analysis and discovery algorithms with the objective of finding unsuspected relationships and summarizing data in ways that are both understandable and useful. When large chunks of data are analyzed in real-time, it points out what is roaming in a network and, based on this, leads to detections and findings in areas of intrusion and vulnerability. Data mining is just concerned with collecting, housing, and analyzing data. Rules and definitions have to act upon this dataset to confirm and further categorize data so that it houses even more meaning and value.
- How can we effectively test and implement an IDS using a Naïve Bayes Classification technique?
- Naïve Bayes Classification can be implemented in multiple ways. It can be slotted on top of SQL query or on data warehouses where the algorithm continuously cross-references in real-time with its definitions to what is presented to classify logs so that administrators have a better chance of detecting anomalies and fluctuations in the network. This simply acts as an early warning mechanism and not a cure.

11.4.2 Findings

Data mining helps in discovering patterns both visible and hidden, which can further be analyzed to better understand data. Real-time data is a requirement to remain relevant in today's market and must be provided through advanced electronic communications technologies such as digital signage. Gathering and analyzing real-time data using data mining techniques enhances an organization's foundation and survivability in today's market. This is very much applicable to networks and infrastructure as every node and element in an organization is connected to a network.

Having control and understanding of your network movements and traffic is essential to determine what is happening on your network and to ascertain whether network traffic is authentic or not and whether authorized individuals are actually accessing the systems or are they unauthorized users. All this is possible through collecting real-time data on a network and cross-referencing them against set user lists and privileges. Naïve Bayes Classification is impeccable in this context as it gives you a probability of what can be classified as an anomaly or what can be considered a usual event. Again, this is just a probability, and every probability is as good as its algorithm in terms of predefined datasets and classifiers. This is called the data schema of the algorithm, which is basically its knowledge set. This is the cross-referencing data to which real-time traffic and movement are referenced against to obtain hits or faults.

11.5 DISCUSSION

The findings from this study give a comparative study of anomaly detection schemes for identifying novel network intrusion detection techniques. Experimental results have been gathered and analyzed, and the Naïve Bayes classifier model is seen to surpass detections of unusual activities and network intrusions. Our experiment is surrounded by the KDD dataset analysis used in IDS. In our research, we realized that the Naïve Bayes algorithm is based upon conditional probability. That is that it uses the Naïve Bayes theorem for classification purposes. In this case, we have a total number of 213 instances, out of which Naïve Bayes has correctly classified 201 instances (94% of accuracy), and 12 are incorrectly classified (6% of non-accuracy). In our case, some of the diagonal elements of the confusion matrix are 119 added with 82, which equals 201, which means that correctly classified instances are equal to the sum of the diagonal elements of the confusion matrix. Whereas non-diagonal elements sum is equal to incorrectly classified instances.

In the course of our research, few highlighting factors to note was that the KDD dataset had missing details, which were filtered and pre-processed in order to get the right results. Most of the results generated were from the 1999 KDD Cup, as false-positive results were close to what we assumed. The results showed that the intrusions were detected and that it also detected genuine files as intrusions that had no significant threat toward the network.

The major challenge faced in our research was that the anomaly-based detection method creates different levels of hypothesis, which needs different tools to interpret and analyze it correctly. Since we are dealing with information faced across the network, one of the contributing limitations here is that most organizations are not willing to allow data mining done on their data to detect intrusion.

Since it deals with numerical data, there are some uncertainties raised as to how detections are going to be placed and, if detected, what are the measures taken to prevent it in the future.

Since the dataset will be different for all the organizations, the time taken will also be different. For some, it might take seconds, as for others, it might take hours. In addition, the actual dataset can also be hard to integrate into the tool itself as the challenge here is that it is a limited dataset to be known before, but as anomalies are detected, its schema is updated to prevent the same anomalies again.

11.6 CONCLUSION

In this paper, we have proposed a system in which we can use a Data Mining Technique such as the Naïve Bayes Classification algorithm to detect intrusions in the system. Network Intrusion Detection System plays a very important role in detecting intrusions during the flow of data from one network to another network. The information age is transforming our economy toward the digital cyber realm in which everything is connected to each other and can communicate. This gives hackers a larger playing field as new ways are discovered to penetrate networks.

IDS gives organizations an extra layer of detection from internal and external penetrators. Our research looked at the use of the KDD dataset to investigate the anomalies found in the system at any given point in time. We used the application NetBeans to cater for the data mining analysis. However, there is potential for other tools which can deliver the same or better results. This classification technique helps in gathering required data which can be used for further decision-making on how to protect the sensitive information of the organization.

11.7 RECOMMENDATIONS AND FUTURE RESEARCH

For future work on this research, our recommendation is simple, we can use the technique which we used to find anomalies and hidden patterns to step into other fields; for instance, we can look into using either Naïve Bayes or even KNN technique to investigate fraud detection in any field. Usually, this is not being practiced. However, there is potential for this technique to be used by Auditors, Cyber Security Specialists, and Forensic Analysts. This is used in banks for merchant fraud detection but through the use of the Microsoft Access feature.

In addition to the above, we can use Wireshark to detect packets of information and analyze them using Weka or NetBeans for a more thorough data mining outcome. Why we are suggesting Wireshark as the tool of choice for future works is that in Wireshark, we can see the breakdown of the entire OSI (Open Systems Interconnections) model of the network. This is something very challenging, yet if given the time and resources, we can explore this given gray area and use the ethical hacking tool to its full potential.

REFERENCES

Abraham, A., Grosan, C., & Chen, Y. (2005). Cyber Security and the Evolution of Intrusion Detection Systems. *Information Management & Computer Security—IMCS*, 1(1), 74–82.

Arvidson, M., & Carlbark, M. (2003). Intrusion Detection Systems: Technologies, Weaknesses and Trends (3390 Student Thesis). Institutionen för Systemteknik. Retrieved from http://urn.kb.se/resolve?urn=urn:nbn:se:liu:diva-1614.

Bace, R. G. (2000). *Intrusion Detection*. Macmillan Technical Publishing.

Barot, V., & Toshniwal, D. (2012). *A New Data Mining Based Hybrid Network Intrusion Detection Model. Paper Presented at the 2012 International Conference on Data Science & Engineering (ICDSE)*.

Basak, A. (2014). *A Hybrid Intrusion Detection Model Using*. Academia.

Bolatiki, M. (2018). Digital Fiji to Revolutionize Way We Access Govt Information Services [Press Release]. Retrieved from https://fijisun.com.fj/2018/06/15/digitalfiji-torevolutionise-way-we-access-govt-info-services/

Chai, H., Lei, J., & Fang, M. (2017). Estimating Bayesian Networks Parameters Using EM and Gibbs Sampling. *Procedia Computer Science, 111*, 160–166.

Chary, K. C. (2012). Data Mining, Intrusion Detection System—A Study. *International Journal of Advanced Research in Computer Science, 3*(1), 434–437.

Chiche, A., & Meshesha, M. (2017). Constructing a Predictive Model for an Intelligent Network Intrusion Detection. *International Journal of Computer Science and Information Security, 15*(3), 392.

Deng, Y., Liu, Y., & Zhou, D. (2015). An Improved Genetic Algorithm with Initial Population Strategy for Symmetric TSP. *Mathematical Problems in Engineering, 2015*, 6.

Farid, D. M. (2010). Combining Naive Bayes and Decision Tree for Adaptive Intrusion Detection. *arXiv preprint arXiv: 1005.4496*.

Goeschel, K. (2016). *Reducing false positives in intrusion detection systems using data-mining techniques utilizing support vector machines, decision trees, and Naive Bayes for off-line analysis. Paper presented at the SoutheastCon 2016*.

Govindarajan, M., & Chandrasekaran, R. M. (2011). Intrusion Detection Using Neural Based Hybrid Classification Methods. *Computer Networks, 55*(8), 1662–1671.

Gujar, S. S., & Patil, B. M. (2014). Intrusion Detection Using Naïve Bayes for Real-Time Data. *International Journal of Advances in Engineering & Technology, 7*(2), 568–574.

Karthick, R. R., Hattiwale, V. P., & Ravindran, B. (2012). *Adaptive Network Intrusion Detection System Using a Hybrid Approach. Paper presented at the 2012 Fourth International Conference on Communication Systems and Networks (COMSNETS 2012)*.

Lee, W., & Stolfo, S. J. (2000). A Framework for Constructing Features and Models for Intrusion Detection Systems. *ACM Transactions on Information and System Security, 3*(4), 227–261.

Malik, A. J., Shahzad, W., & Khan, F. A. (2015). Network Intrusion Detection Using Hybrid Binary PSO and Random Forests Algorithm. *Security and Communication Networks, 8*(16), 2646–2660.

Markou, M., & Singh, S. (2003). Novelty Detection: A Review—Part 1: Statistical Approaches. *Signal Processing, 83*(12), 2481–2497.

Mukherjee, S., & Sharma, N. (2012). Intrusion Detection Using Naive Bayes Classifier with Feature Reduction. *Procedia Technology, 4*, 119–128.

Nielsen, T. D., & Jensen, F. V. (2013). *Bayesian Networks and Decision Graphs*. New York: Springer.

Panda, M. (2010). *Discriminative Multinomial Naive Bayes*. Academia.

Panda, M., & Patra, M. (2009). Evaluating Machine Learning Algorithms for Detecting Network Intrusions. *International Journal of Recent Trends in Engineering, 1*(1), 472.

Rahul, V. K., Vinayakumar, R., Soman, K. P., & Poornachandran, P. (2018). *Evaluating Shallow and Deep Neural Networks for Network Intrusion Detection Systems in Cyber Security. 2018 Ninth International Conference on Computing, Communication and Networking Technologies (ICCCNT)* (pp. 1–6). Piscataway, NJ: The Institute of Electrical and Electronics Engineers (IEEE).

Rao, U. H., & Nayak, U. (2014). Intrusion Detection and Prevention Systems. *The InfoSec Handbook: An Introduction to Information Security* (pp. 225–243). Berkeley, CA: Apress.

Salo, F., Injadat, M., Nassif, A., Shami, A., & Essex, A. (2018). Data Mining Techniques in Intrusion Detection Systems: A Systematic Literature Review. *IEEE Access*, *6*, 56046–56058.

Sethi, K. K., Mishra, D. K., Solanki, G., & Mishra, B. (2009). *Key Issues of Security and Integrity in Third Party Association Rule Mining. Paper presented at the Proceedings of the 2009 Second International Conference on Emerging Trends in Engineering & Technology.*

Shyara Taruna, R, & Hiranwal, S (2013). Detecting Intrusion in Data Mining Using Naive Bayes Algorithm. *International Journal for Scientific Research and Development*, *1*(9), 1759–1762.

Sisodia, M. S., Sharma, S. K., & Pandey, P. (2012). Anomaly Based Network Intrusion Detection by using Data Mining. *International Journal of Advanced Research in Computer Science and Electronics Engineering*, *1*(1), 33–38.

Subbalakshmi, G., Ramesh, K., & Rao, M. (2011). Decision Support in Heart Disease Prediction System using Naive Bayes.

Translated by ContentEngine LLC (2019). Symantec Supports Banking in Foreseeing Cyber Attacks. CE Noticias Financieras. Retrieved from http://ezproxy.usp.ac.fj/login?url= https://search.proquest.com/docview/2186330569?accountid=28103

Tsuruoka, Y., & Tsujii, J. I. (2003). *Training a Naive Bayes Classifier via the EM Algorithm with a Class Distribution Constraint. Paper presented at the Proceedings of the Seventh Conference on Natural Language Learning at HLT-NAACL 2003—Volume 4*, Edmonton, Canada.

Wahl, R. S. (2016). Latency in Intrusion Detection Systems (IDS) and Cyber-Attacks: A Quantitative Comparative Study (10132049 Ph.D.). Capella University, Ann Arbor. Retrieved from http://ezproxy.usp.ac.fj/login?url=https://search.proquest.com/docview/1810428478?accountid=28103.

Warrender, C., Forrest, S., & Pearlmutter, B. (1999). *Detecting Intrusions Using System Calls: Alternative Data Models. Paper presented at the Proceedings of the 1999 IEEE Symposium on Security and Privacy (Cat. No.99CB36344).*

Zhang, B., Liu, Z., Jia, Y., Ren, J., & Zhao, X. (2018). Network Intrusion Detection Method Based on PCA and Bayes Algorithm. *Security and Communication Networks*, 2018, 11.

12 Data Security Analysis in Mobile Cloud Computing for Cyber Security

Sam Goundar

RMIT University, Vietnam

Akashdeep Bhardwaj

University of Petroleum and Energy Studies, Dehradun, India

CONTENTS

12.1 INTRODUCTION

According to O'Dea (2020), "the number of mobile devices operating worldwide is expected to reach 17.72 billion by 2024, an increase of 3.7 billion devices compared to 2020 levels". This is an indication of how pervasive mobile devices have become and how dependent we have become on using mobile devices in our daily lives. However, we are just getting started. With the anticipation of 5G networks rolling out and connecting our mobile devices, making connectivity faster and ubiquitous, and one can only imagine the uses of our mobile devices. The reason why we are able to use our mobile devices like our laptops and desktops is because of the convergence of mobile technologies with cloud computing technologies. Mobile cloud computing services are now being used everywhere by everyone for everything. Whether it is for checking your emails, banking, and logging into your courses, watching videos online, or chatting on social media, all these activities and services need to be secured in cyberspace (cyber-security) as you do not want anyone else reading your emails or logging into your bank account.

This chapter focuses on the security challenges and risks we face with mobile cloud computing when we use our mobile devices. Because of the convergence of two different technologies in mobile cloud computing, the inherent security risks of each technology are two-pronged for the mobile device user. Not only do they have to deal with the security risks of cloud computing technology, but mobile technology as well. For example, (Bhardwaj & Goundar, 2020) indicate that the "challenges and risks faced in cloud security services are in the areas which include identity access management, web security, email security, network security, encryption, information security, intrusion management, and disaster management while implementing a cloud service infrastructure." While data leakage, compromised privacy, unauthorized access, phishing attacks, spyware attacks, network spoofing attacks, surveillance attacks, etc., have been reported by (Mishra & Thakur, 2019) in their paper titled "A Survey on Mobile Security Issues."

In addition to the above, the human error of losing mobile devices cannot be mitigated with any technological solutions. Although applications like "find my phone" exist as well as other locators and GPS, we also have an active community of users in the society that are able to overcome that and break into the device to get data through encryption. Then the issue of readily available free apps for downloads makes the users vulnerable to unsafe malware apps that not only spy on the users but also extract data from them. Many mobile users remain unprotected from malware.

Over the past few years, computing technology is evolving at an extensively rapid rate, and the demand for storing data is reaching its peak. Mobile cloud computing is also one of the emerging storage technological transformations. Mobile devices, for instance, tablets and smartphones are now the new age necessity and are not restricted by location and time; thus, the progress of mobile computing has become an influential tendency in ICT. On the other hand, cloud storage is considered to be the new generation computing infrastructure providing various platforms for end-users. Cloud storage also is the solution to storing data from mobile computing. This is known as mobile cloud computing. Mobile cloud computing is bringing in a revolution in terms of processing and storing capabilities of data. However, is the data given in by the end-users safe and secured?

> Mobile cloud computing at its simplest refers to an infrastructure where both the data storage and data processing happen outside of the mobile device. Mobile cloud applications move the computing power and data storage away from mobile phones and into the cloud, bringing applications and mobile computing to not just smartphone users but a much broader range of mobile subscribers.
>
> (Dinh et al., 2011)

Mobile cloud computing initially emerged from cloud computing and mobile computing in order to provide cloud services to mobile users. According to several research, the popularity of mobile cloud computing is increasing massively; hence there is an increased need for authentic security of data stored. This research paper emphasizes data security encounters in cloud computing and suggestions to overcome them. The objectives of this research paper are to critically evaluate mobile cloud computing, understand and explain data security encounters in mobile cloud computing, assess the influence of mobile cloud computing to practical applications of information systems, and suggest possible solutions to the encounters.

Personal mobile devices not properly secured and used can also become a cyber-security risk for organizations and educational institutions. As employees and students take their personal mobile devices to their work and institutions (Bring Your Own Device—B.Y.O.D), they connect to the corporate network and can become the vulnerable link that hackers can use to gain access. Most organizations allow such connections as the employees use these devices for work purposes as well. The number of users on mobile banking platforms has increased dramatically. Almost all banks provide m-Banking and an app for their customers. This has gained attention with hackers who are now developing techniques to intercept banking sessions for monetary gains. The fact that mobile devices connect wirelessly to wi-fi hot spots and routers are also a cause for security breaches. WIFI connections are easier to hack and intercept when compared to wired connections.

12.2   MOBILE CLOUD COMPUTING AND ITS CHALLENGES

Mobile cloud computing is a rich computation of resources bought down to mobile users via a blend of cloud computing, mobile computing, and wireless networks. "Mobile cloud computing promises several benefits such as extra battery life and storage, scalability, and reliability" (Noor et al., 2018).

> However, there are still challenges that must be addressed in order to enable the ubiquitous deployment and adoption of mobile cloud computing. Some of these challenges include security, privacy and trust, bandwidth and data transfer, data management and synchronization, energy efficiency, and heterogeneity.

Noor et al. (2018) further identified "the following research challenges, namely security, privacy and trust, bandwidth and data transfer, data management and synchronization, energy efficiency, and heterogeneity."

Security issues concern both the end-users and cloud service providers because a third party may take part in misusing confidential data that can be malicious (Rahimi et al., 2014). A security vulnerability may occur in numerous issues; thus, our criteria in defining the challenges are highlighted by data security issues in mobile cloud computing, i.e., unreliable authentication and authorization. Authentication is significant while accessing significant information such as bank details, personal credentials, and other confidential information. Authentication is conveyed bidirectional, i.e., the mobile user ought to authenticate to the cloud server, known as mobile-to-cloud authentication, and the cloud server is ought to authenticate to the mobile user, which is known as cloud-to-mobile authentication as stated by Wang, Chen, & Wang (2015).

Using restraint manuscript imputations, end-users are apt to use the simplest of the passwords and tend to use the same password for multiple applications, which leads to mobile applications being more reluctant to authentication threats. Al-Muhtadi et al. (2011), in their paper, dissected that "mobile-to-cloud authentication structure for universal computing surroundings using wearable devices and biometrics will enable an enhanced protection for authenticating a mobile client's identity than only password-protected systems." Then authentication threat leads to authorization. Authorizing an application software to give access to a user's data without issuing the client's credentials is another challenge. This is mostly seen among social media applications. Several researchers concluded that the access to sign-in possibility and timeout information for authorization is reluctant to copying attacks; thus, if another party is entitled to a copy will be authorized to accessing the client's data. Authorization thus needs to be considered at a granular level based on authentication.

According to (Noor et al., 2018), "the mobile cloud computing environment is distributed and dynamic." "Therefore, it is crucial that any proposed mobile cloud computing architecture should be scalable. Scalability means that the mobile cloud computing architecture has the ability to grow and shrink by increasing or decreasing the resources (and users) respectively". "Mobile cloud computing architectures that follow a centralized architecture design will suffer from a number of issues such as scalability and security" (Figure 12.1).

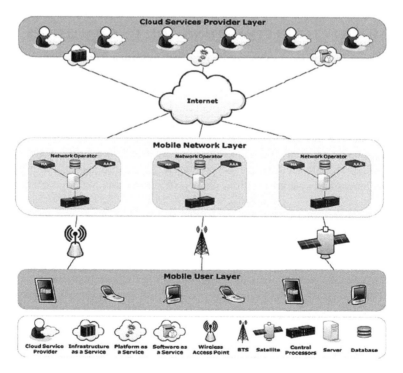

FIGURE 12.1   Architecture of mobile cloud computing (Noor et al., 2018).

Other issues and challenges of mobile cloud computing, as mentioned by Aliyu et al. (2020), are energy efficiency, security and privacy, user satisfaction with mobile apps, and quality of service. As mobile devices offload all computation for processing to near nodes, some of the factors affecting offloading decisions include network data cost, cloud service cost, data privacy, energy efficiency, and execution support. Success offloading requires mobile device resources, good network connectivity, adequate network bandwidth, and cloud resources.

12.3   LITERATURE REVIEW

An intelligent malware detection system called (MOBDroid) is proposed by (Ogwara, Petrova, & Yang, 2020) to protect end-user mobile devices that use the Android Operating System. The MOBDroid system is a permission-based security system that can be installed on Android mobile devices and will detect malware within the mobile cloud computing network. After studying the unique permissions requested by 28,306 malicious apps from the mobile devices (Ogwara, Petrova, & Yang, 2020), built a library of these permissions to detect malicious apps based on the permission requests. "In the experiments conducted, we obtained classification accuracy of 96.89%, a detection rate of 98.65%, and false-negative rate of 1.35%". With an almost 98% detection rate for malware on mobile devices, the MOBDroid could be an ideal solution for data security concerns, issues, and challenges of mobile cloud computing.

To "strengthen the security of access control protocols for mobile cloud environment," Agrawal and Tapaswi (2019) used dynamic attributes of mobile devices. They state that "the weak or disconnection issue of the mobile network is a critical task to deal with." Their "proposed approach provides access control as well as data confidentiality using dynamic attributes of encryption. They used "pairs of mobile agents to deal with the issue of network connection." The technique used is to distribute the secret key using the anonymous key-issuing protocol, and this ensures that the user remains unknown. They implemented this approach in a real mobile cloud computing environment and evaluated its performance under various parameters.

"Cloud resources can help in overcoming the memory, energy, and other computing resource limitations of mobile devices," as stated by Vemulapalli, Madria, and Linderman (2020). The authors add, "the mobile cloud computing applications can address some of the resource constraint issues by offloading tasks to cloud servers." However, "despite these advantages, mobile cloud computing is still not widely adopted due to various challenges associated with security in mobile cloud computing framework including issues of privacy, access control, service level agreements, interoperability, charging model, etc.". In their book chapter, they "focus on the challenges associated with security in mobile cloud computing, and key features required in a security framework for mobile cloud computing." Initially, they "describe key architectures pertaining to various applications of mobile cloud computing," and later, they "discuss few security frameworks proposed for mobile cloud computing" in terms of handling privacy, security, and attacks.

According to Shahzad and Hussain (2013), cloud computing delivers services and resources to users via a public network. The ownership of the infrastructure is majorly held by the third parties known as cloud service providers. They elaborate that there are three cloud service models, i.e., infrastructure as a Service (IaaS), Platform as a Service (PaaS), Software as a Service (SaaS), and three cloud deployment methods, i.e., Private cloud, Public cloud, and Hybrid cloud. Moreover, the authors discuss that mobile cloud computing emerged from cloud computing, whereby it is a combination of cloud computing and mobile computing, inviting the attention of an increased number of mobile device users and industrialists. Mobile cloud computing uses the storage capabilities of the cloud to reduce issues of mobile users such as slow processing power, limited battery power, low internet bandwidth, and small storage space.

Furthermore, (Patel, 2018) emphasizes the issues and challenges of mobile cloud computing. Firstly, data security and privacy issues, he lists associated risks, i.e., data theft risk., the privacy of data belongs to customers, violation of privacy rights, loss of physical security, handling of encryption and decryption keys, security and auditing issues of virtual machines, lack of a standard to ensure data integrity, services incompatibility because of different vendors involvement. Secondly, architecture and cloud service delivery models issue, he lists some issues, i.e., computing offloading, security for mobile users/applications/data, improvement in efficiency rate of data access, the context-aware mobile cloud services, migration and interoperability, service level agreement, cost and pricing. Mobile cloud infrastructures issues (Akherfi, Gerndt, & Harroud, 2018), i.e., attacks on virtual machines, vulnerabilities that exist at platform level, phishing, authorization and authentication, attacks from local users, hybrid cloud security management issues. Lastly, mobile cloud communication channels issues (Carreiro & Oliveira, 2019), i.e., access control attacks, data integrity attacks, attacks on authentication, attacks on availability.

Additionally, Almaiah and Al-Khasawneh (2020) discusses some advantages of mobile cloud computing, i.e., long battery life, enhanced processing power, and data storage space, more data and application reliability, scalability, multi-tenancy, and flexible Integration. The author also elaborates on applications of mobile cloud computing, i.e., mobile commerce, mobile learning, mobile healthcare, and mobile gaming. The author concludes that mobile cloud computing is the new generation technology providing mobile device users services and resources, but it is associated with vast data security risks and mobile service providers need to improve their security technologies to minimize data security concerns.

12.4   RESEARCH METHODOLOGY

The basic and applied research can be quantitative or qualitative, or even both (Goundar, 2013). Quantitative research is based on the measurement of quantity or amount. Here a process is expressed or described in terms of one or more quantities. Qualitative research is concerned with qualitative phenomena involving quality. It is non-numerical, descriptive, applies reasoning, and uses words. Its aim is to get the meaning, feeling and describe the situation. The decision to use either the quantitative method or the qualitative method depends on the type and volume of data that you intend to collect and how the data will be analyzed. It also depends on whom you are going to collect the data from and how you are going to collect that data. As we intend to collect data from mobile device users and analyses data security issues for cyber-security, the quantitative method was more suited to this research.

For this research, and because of time constraints and resources, we have used survey questionnaires to collect primary data. Survey questionnaires are designed to gather brief information and are relatively easy for the participants as they can just make selections for their responses. Surveys are capable of obtaining information from large samples of the population (Glasow, 2005). They are also well suited to gathering demographic data that describe the composition of the sample (McIntyre, 1999). Surveys are inclusive in the types and number of variables that can be studied, require minimal investment to develop and administer, and are relatively easy for making generalizations (Bell, 1996). Surveys can also elicit information about attitudes that are otherwise difficult to measure using observational techniques (McIntyre, 1999). Surveys can also be used to assess needs, evaluate demand, and examine the impact (Salant & Dillman, 1994).

As researchers in the field of computer science and information technology, the decision to use online survey questionnaires was appropriate. And the use of the free subscription of Survey Monkey was to avoid unnecessary costs for the research. Online data collection has become the norm. We can hardly see researchers printing survey questionnaires and posting them to participants. And why would the participants bother to post it back? According to Topp & Pawloski (2002), "online data collection is becoming an essential and efficient tool for evaluators, researchers, and other educators." A study conducted by Sax, Gilmartin, and Bryant (2003) used four different survey modes (paper-only, paper with web option, web-only with response incentive, and web-only without response incentive) to gauge response rates and nonresponse bias. The response rates from online survey questionnaires were better than other modes.

12.4.1 PARTICIPANTS

A total of 168 participants responded to the survey questionnaire, 55% of the respondents were between the ages of 18–25 who were either undergraduate or graduate students within the information systems program. The other 45% of participants were working within the IT industry or dealing with information systems in their organizations. From those working in the industry, 8% were between the ages of 26–33, 5% between the ages of 34–39, and 4% were 40 and above. All participants had a university education and understood the concept of data security as related to their mobile devices. All participants volunteered to participate freely and were not provided any incentives to participate.

12.4.2 MATERIALS

The survey was carried out using questionnaires answered by the participants via an online survey platform, namely, Survey Monkey, with the participants being sent the following link to take part in the survey: (https://www.surveymonkey.com/r/7 SFNY5W) the survey consisted of 20 questions based on multiple choices and Likert scale type questions. During the pilot survey questionnaires testing, it was estimated that the survey could be completed in 5 minutes as the participant basically had to make selections provided based on each question. The participants were asked a question relating to their personal experiences with data security issues with their mobile devices. According to Varela et al. (2016), "Survey Monkey has provided some positive aspects, such as: easier access, avoidance of input and data coding errors, a faster distribution and saving time and cost."

12.4.3 PROCEDURES

The online participants were requested to give in their email address, occupation, education level, and their experiences with data security in mobile cloud computing. Only the participants that worked within the IT industry or students doing computer science and information systems majors were sent the initial invitation via emails. Upon their agreement to volunteer to participate and emailed consent, the link to the survey questionnaires were sent to them. The survey was open for 60 days. A few email reminders were sent on a fortnightly basis. The responses were then downloaded and thoroughly analyzed. With the use of these analyses, results were generated.

12.4.4 RESULTS

Survey Question: What are the Contributing Factors preventing Mobile Users not to opt for Mobile Cloud Computing? (Figure 12.2)

Survey Question: Have You Ever Encountered Data Security Issues While Using Mobile Cloud Computing? (Figure 12.3)

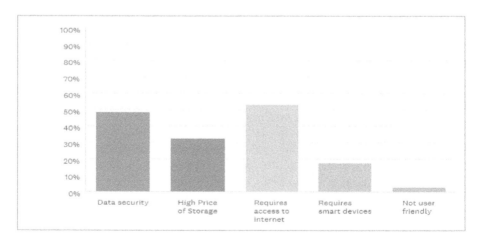

FIGURE 12.2 Contributing factors stopping mobile users not to opt for mobile cloud computing.

Source: IS413_201803: Research Questionnaire.

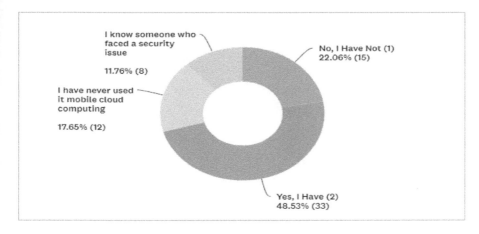

FIGURE 12.3   Have you ever encountered data security issues while using mobile cloud computing?

Source: IS413_201803: Research Questionnaire.

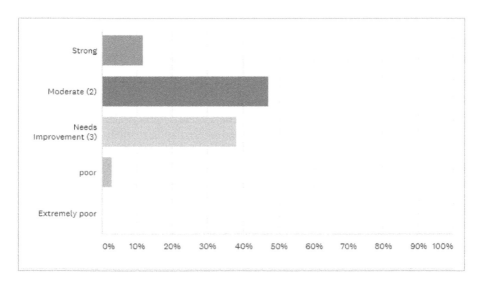

FIGURE 12.4   What do you think of mobile cloud computing's data security measures?

Source: IS413_201803: Research Questionnaire.

Survey Question: What Do You Think of Mobile Cloud Computing's Data Security Measures? (Figure 12.4)

Survey Question: Which of the Following Do You Prefer? (Figure 12.5)

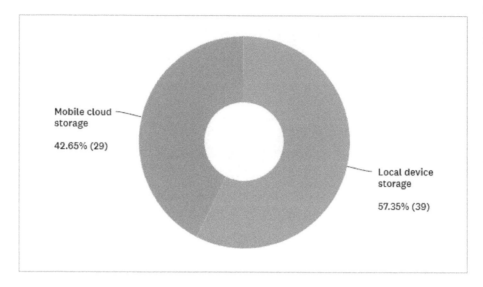

FIGURE 12.5   Which of the following do you prefer?

Source: IS413_201803: Research Questionnaire.

12.5   DATA SECURITY IN MOBILE CLOUD COMPUTING

A Secure mobile user-based Data Service Mechanism (SDSM) was designed by Jia et al. (2011)) to enable a granular level of access control to user's data in the cloud. After reviewing and evaluating a number of data security proposals, models, frameworks, and mechanisms for mobile cloud computing, the researchers could ascertain that none of them were suitable and comprehensive enough for data security in mobile cloud computing. One of the reasons behind the shortcomings of other proposals was the fact that it did to take into consideration the dynamic nature of mobile device users on the move. The mobile device users join and leave the mobile cloud computing networks haphazardly. Because of this erratic mobile device user behavior and the inability to monitor them for long, a mechanism was needed to ensure that any mobile device user that joined would only have granular access to the network and would only be able to access their own data to ensure confidentiality. The mechanism achieves confidentiality and granular level access to data by outsourcing not only the data but also the security management to the mobile cloud for trust.

Mobile cloud apps need to interact in real-time with several other resources and nodes within the mobile cloud computing network leading to many issues of security and privacy. In a research paper published by Tawalbeh et al. (2017), the main issues of mobile cloud computing security issues were reviewed and collated. The main issues were found to be security, performance, and quality, according to Tawalbeh et al. (2017). A cloudlet-based mobile cloud computing framework for data security that could be securely implemented was proposed by the researchers. The prototype that was developed and tested based on this framework used the trust delegation technique to provide data security better than the traditional frameworks. The performance was better, latency reduced, and throughput increased with the cloudlet-based mobile cloud computing framework. Additionally, it was observed that the quality of service improved, the architecture was scalable and available with the cloudlet model.

Within the last two years, there have been some noteworthy data security proposals in mobile cloud computing as well. For example, "Homomorphic Encryption as a Service for Outsourced Images in Mobile Cloud Computing Environment" was proposed by Ibtihal and Hassan (2020). While Lo'ai and Saldamli (2019) writes about "Reconsidering big data security and privacy in cloud and mobile cloud systems." "A trustworthy agent-based encrypted access control method for mobile cloud computing environment" was implemented and tested by Agrawal and Tapaswi (2019).

12.6 DISCUSSION

The survey results indicate that a majority of participants prefer local device storage over mobile cloud storage. Some of the reasons behind these odd statistics can be they have either faced data security issues in cloud storage or have known participants who have incurred great losses as a result of using mobile cloud computing. When asked to rate the data security measures of mobile cloud computing, a majority of the participants stated it needs improvement rather than strongly support its usage. This also could be for the same reason as to why they prefer local storage over mobile cloud storage. According to the results, a large number of participants convey the reason for not using mobile cloud computing is due to data security issues. Some other reasons were the high price of storage, requires internet access, and requires smart devices, and not a friendly user interface. As per our survey, it is a clear indicator that the data security measures are the most likely reason stopping the expected number of participants from using mobile cloud computing.

A novel framework is proposed by Elgendy et al. (2018) to take advantage of the security issues of the cloud by offloading all intensive computations to the cloud. Their "framework uses an optimization model to determine the offloading decision dynamically based on four main parameters, namely, energy consumption, CPU utilization, execution time, and memory usage." They also suggest "a new security layer is provided to protect the transferred data in the cloud from any attack."

12.7 CONCLUSION

To conclude, the data security of mobile cloud computing is not reliable enough to secure the data of the end-users, and therefore possible measures should be taken by the respective service providers to minimize security encounters in mobile cloud computing. Authentication, authorization, and accountability are the most significant factors in order to have reliable, safe, and secured data processing and storage. There is more scope of research to be conducted regarding solutions to combat the encounters in mobile cloud computing security issues. Data security and privacy still remain major issues for mobile device users. The violation of privacy and trust on social media platforms have left mobile device users worried about using mobile cloud computing. Research, models, frameworks, and mechanisms that allow proper data security in mobile cloud computing are needed to reassure mobile device users that they can safely compute with mobile cloud computing and that there will be no data security issues.

REFERENCES

Agrawal, N., & Tapaswi, S. (2019). A trustworthy agent-based encrypted access control method for mobile cloud computing environment. *Pervasive and Mobile Computing*, *52*, 13–28.

Akherfi, K., Gerndt, M., & Harroud, H. (2018). Mobile cloud computing for computation offloading: issues and challenges. *Applied Computing and Informatics*, *14*(1), 1–16.

Aliyu, A., Abdullah, A. H., Kaiwartya, O., Hussain Madni, S. H., Joda, U. M., Ado, A., & Tayyab, M. (2020). Mobile cloud computing: taxonomy and challenges. *Journal of Computer Networks and Communications*, 2020.

Almaiah, M. A., & Al-Khasawneh, A. (2020). Investigating the main determinants of mobile cloud computing adoption in university campus. *Education and Information Technologies*, *25*(4), 3087–3107.

Bell, S. (1996). *Learning with information systems: learning cycles in information systems development*. New York: Routledge.

Bhardwaj, A., & Goundar, S. (2020). Cloud computing security services to mitigate DDoS attacks. *Cloud computing security-concepts and practice*. London: IntechOpen.

Carreiro, H., & Oliveira, T. (2019). Impact of transformational leadership on the diffusion of innovation in firms: application to mobile cloud computing. *Computers in Industry*, *107*, 104–113.

Dinh, H., Lee, C., Niyato, D., & Wang, P. (2011). A survey of mobile cloud computing: architecture, applications, and approaches. *Wireless Communications and Mobile Computing*, *13*(18), 1587–1611.

Elgendy, I., Zhang, W., Liu, C., & Hsu, C. H. (2018). An efficient and secured framework for mobile cloud computing. *IEEE Transactions on Cloud Computing*, *9*(1), 79–87.

Glasow, P. A. (2005). Fundamentals of survey research methodology. Retrieved August, 26, 2021, from http://www.uky.edu/~kdbrad2/EPE619/Handouts/SurveyResearchReading.pdf

Goundar, S. (2013). *Research methodology and research method*. Victoria University of Wellington. https://www.researchgate.net/publication/333015026_Chapter_3_-_Research_Methodology_and_Research_Method

Ibtihal, M., & Hassan, N. (2020). Homomorphic encryption as a service for outsourced images in mobile cloud computing environment. In *Cryptography: breakthroughs in research and practice* (pp. 316–330). Hershey, PA: IGI Global.

Jia, W., Zhu, H., Cao, Z., Wei, L., & Lin, X. (2011, April). SDSM: a secure data service mechanism in mobile cloud computing. In *2011 IEEE conference on computer communications workshops (infocom wkshps)* (pp. 1060–1065). New York: IEEE.

Lo'ai, A. T., & Saldamli, G. (2019). Reconsidering big data security and privacy in cloud and mobile cloud systems. *Journal of King Saud University—Computer and Information Sciences*, *33*(7), 810–819.

McIntyre, L. J. (1999). *The practical skeptic: core concepts in sociology*. Mountain View, CA: Mayfield Publishing.

Mishra, S., & Thakur, A. (2019). A survey on mobile security issues. *Proceedings of Recent Advances in Interdisciplinary Trends in Engineering & Applications (RAITEA)*.

Noor, T. H., Zeadally, S., Alfazi, A., & Sheng, Q. Z. (2018). Mobile cloud computing: challenges and future research directions. *Journal of Network and Computer Applications*, *115*, 70–85.

O'Dea, S. (2020). Number of mobile devices worldwide 2020–2024. Retrieved February 16, 2021, from https://www.statista.com/statistics/245501/multiple-mobile-device-ownership-worldwide/

Ogwara, N. O., Petrova, K., & Yang, M. L. B. (2020). MOBDroid: an intelligent malware detection system for improved data security in mobile cloud computing environments. In *2020 30th International Telecommunication Networks and Applications Conference (ITNAC)* (pp. 1–6). New York: IEEE.

Patel, R. (2018). Survey on mobile cloud computing. *International Journal for Technological Research in Engineering*, 5(8).

Qayyum, R., & Ejaz, H. (2020). Data security in mobile cloud computing: a state of the art review. *International Journal of Modern Education & Computer Science*, *12*(2), 30–35.

Rahimi, M. R., Ren, J., Liu, C. H., Vasilakos, A. V., & Venkatasubramanian, N. (2014). Mobile cloud computing: a survey, state of art and future directions. *Mobile Networks and Applications*, *19*(2), 133–143.

Salant, P., & Dillman, D. A. (1994). *How to conduct your own survey*. New York: John Wiley and Sons.

Sax, L. J., Gilmartin, S. K., & Bryant, A. N. (2003). Assessing response rates and nonresponse bias in web and paper surveys. *Research in Higher Education*, *44*(4), 409–432.

Shahzad, A., & Hussain, M. (2013). Security issues and challenges of mobile cloud computing. *International Journal of Grid and Distributed Computing*, *6*(6), 37–50.

Tawalbeh, L. A. A., Ababneh, F., Jararweh, Y., & AlDosari, F. (2017). Trust delegation-based secure mobile cloud computing framework. *International Journal of Information and Computer Security*, *9*(1–2), 36–48.

Topp, N. W., & Pawloski, B. (2002). Online data collection. *Journal of Science Education and Technology*, *11*(2), 173–178.

Varela, C., Ruiz, J., Andrés, A., Roy, R., Fusté, A., & Saldaña, C. (2016). Advantages and disadvantages of using the website SurveyMonkey in a real study: psychopathological profile in people with normal-weight, overweight and obesity in a community sample. *E-Methodology*, *2016*(3), 77–89.

Vemulapalli, C., Madria, S. K., & Linderman, M. (2020). Security frameworks in mobile cloud computing. In *Handbook of computer networks and cyber security* (pp. 1–41). Springer, Cham.

Wang, Y., Chen, R., & Wang, D. C. (2015). A survey of mobile cloud computing applications—perspectives and challenges. *Wireless Personal Communications*, *80*, 1607–1623.

13 A Comprehensive Review of Investigations of Suspects of Cyber Crimes

Nehinbe Joshua Ojo
Federal University, Oye, Nigeria

CONTENTS

13.1 INTRODUCTION

Cyber crimes are increasingly generating economic and political impacts on global technologies that are designed to solve human problems on a daily basis (CSIS, 2018). Apart from the disastrous and retrogressive impacts of cyber crimes on the global economy, as technologies are more smatter than ever before, the procedures for interrogating suspects of cyber crimes are coming up with series of disputable

DOI: 10.1201/9781003140023-13

issues in cyber security and forensics in recent years. Accordingly, an individual, institution, or company may be sued or charged to a court of law for wrongly accusing their fellow human being or competitors. According to Human Rights Best Practices (HRBP), interrogators of suspects are not expected to torture or subject them to poor treatment in the course of gathering evidence.

These new dilemmas have raised two grave concerns that are centrally important in the interrogations of suspects of cyber crimes. But then, the complexity of these problems would have been lessened if past studies on interrogations of suspects of cyber crimes in organizations have properly categorized suspects of cyber crimes. Meanwhile, significant numbers of the above studies have not equally clarified how suspects of cyber crimes should be managed or treated before recourse to the law enforcement agencies such as Police, Criminal Justice, and FBI for interrogations (Layton, 2020). Basically, cyber crimes are illegal activities involving users (intruders) and computer networks such as mobile devices and the Internet. Legally, interrogators should not detain any suspect of cyber crimes beyond 24 hours without charging him or her to court in most countries. Most constitutions and national laws have prohibited torture, brutal, merciless, or humiliating treatment of suspects of crimes in general. The law also prescribes series of penalties for persons and law enforcement agents that violate anti-torturing laws either by aids and abetting or by any act of committing blunder relating to torturing suspects of crimes. Nevertheless, some suspects may not be willing to confess without compulsion, coercion, and threats. Thus, system auditors and operators of IDSs in corporate settings are often faced with various challenges whenever the logs of IDSs have been analyzed and suspects of cyber crimes have been readily apprehended. The above forensic issues have raised several legal and technical questions that empirical studies have not answered over the years. For instance, what if the depth of involvement of the person thought to be guilty of a cyber crime is not enough to report him/her to a suitable law enforcement agency? Given the fact that suspects of cyber crimes may be susceptible to wrong accusations for a number of reasons, what if some specialized Trojans, malware, or computer viruses that infected and adopted the computer systems of the suspects were the intruders behind the accusations levied against the suspects? Thirdly, rather than using a sledgehammer to kill a mosquito, what is the recommended level of involvement of a suspect of cyber crimes that is sufficient to litigate or report him/her to the Police? Should organizations that are victims or defendants of cyber crimes always report any suspect apprehended for committing cyber crimes to the Police without conducting any cross-examination of the suspects?

Workplace experience shows that suspects of cyber crimes are not always accused. Thus, organizations must avoid false alarms as much as possible. Unlike Policemen, most corporate culture does not permit organizations to invite and detain relatives of suspects of cyber crimes. One of the strategies to lessen the above challenges is for corporate organizations to deeply confirm every incident and the level of involvement of the suspects in the crime before they will involve any law enforcement agent. Accordingly, a team of investigators comprised of system auditors, forensic accountants, and legal experts drawn from the law, operations, and IT departments should be constituted to interrogate the suspects (Perri and Lichtenwald, 2018).

In Polivanyuk (2019), intruders are prime suspects of cyber crimes. Conceptually, intruders in private and corporate settings may be insiders, outsiders, or connivance. It is plausible that perpetrators that are insiders may be relatives or close associates. Some employees that have access to sensitive information about the operational procedures of the targets or corporation may be accomplices or perpetrators of cyber crime. Contrarily, perpetrators of cyber crimes that are outsiders are certain people that are not close relatives. They are not current employees or members of the organization. In other words, perpetrators of cyber crime in this category may be some of the staff that have disengaged from the organization. More so, perpetrators of cyber crimes that are connivance may signify tacit consent or collusion by two or more allies with the aim to secretly and jointly carry out cyber crimes. The connivance may also decide to divulge useful information about certain targets to their gangsters. Organized crimes are common issues in crime studies. It is an indisputable fact that some connivance may systematically aiding and abetting intruders in the commission of cyber crimes in diverse ways. Hence, interrogators must ascertain at least two things in this respect. They need to ascertain the willingness and unwillingness of the accomplice to secretly participate in cyber crime. Secondly, interrogators need to ascertain the degree of this infringement of the cyber law by establishing various suspects that planned the cyber crime to isolate the possibility of shielding connivance of officials that may be within the interrogation team.

Technically, some suspects are involved in cyber crimes for their roles in inciting intruders. Beyond that, some suspects may be involved in crimes for knowingly or unknowingly divulging corporate information and trade secret to dubious intruders. Crime studies suggest that suspects can passively or actively collaborate with intruders in sharing sensitive information and hacking skills together. For instance, disgruntled employees and employees that were victims of unlawfully disengaged by their employers may be privy to vital security details. It is a well-established fact that some consultants and vendors of third-party's apps (computer applications) may be custodians of significant numbers of sensitive information about their clients. The danger is enormous if any of the above actors should deliberately reveal sensitive information that they keep as trade or professional secret to intruders.

Nonetheless, there is an acute shortage of models for classifying witnesses to cyber crimes that can assist interrogators of suspects of cyber crime in corporate settings over the years. Moreover, significant numbers of past empirical studies have not broadly clarified potential challenges that may extensively influence the acceptability of investigative reports generated during the preliminary investigations the above team have conducted through the lenses of the Police and courts of law (Gehl and Plecas, 2017). Basically, Police have the constitutional responsibility to discontinue and close the cyber case if they find out that it is baseless, a hoax, and a bunch of false accusations.

Another issue is that members of the investigative team in a corporate setting are mainly civilians. By law, civilians are strictly restricted to certain jurisdictions within which their powers to interrogate suspects of cyber crimes can be exercised and prohibited. The locations, society, and religion of the suspects are very important in conducting sound investigations on them to avoid infringement of local laws and their Fundamental Human Rights. Furthermore, circumstances may require the investigators (interrogators) to further engage the suspects for a prolonged period of time. The investigators may wish to cross-examine the suspects at a preliminary stage of the investigations in an unusual time in other to be sure that the organization has a valid case against the suspects. However, the above issues often come with inherent challenges in reality. One of the approaches to lessen these problems is to understand the capability of information that IDS logs can furnish the investigating team. Nehinbe (2011) suggests that significant attributes of IDS logs are less informative for forensic purposes. Useful forensic information can be extracted from logs of IDSs, as shown in Figure 13.1.

Logs of IDSs may not often express all the intentions of intruders, accomplices, and connivance. For these reasons, cyber attacks such as data leakage, blackmailing, attacks designed to evaluate the stronghold of the security measures on corporate networks, and how criminals undertake all their actions may be directly deduced from confessional statements of the suspects rather than exploring IDS logs. Contrarily, sources, destinations, and descriptions of cyber attacks together with their timestamp are veritable information for the investigators to gain insightful evidence on how the cyber attacks originate and how the suspects may be obtained. These three attributes can also help investigators to detect the intentions of the suspects, the aim, the time, and the date of occurrence of the suspicious events.

Unlike the Police that are saddled with the responsibility of investigating crime, the rights that the above team of investigators may possess for them to examine suspects of cyber crimes are clearly limited by law in many countries (Layton, 2020). Consequently, the dilemma is that most of the strategies that Police can adopt (such as torture and inducement by mood-altering drugs) to determine the level of involvement of suspects in cyber crimes are not applicable to the above teams of investigators. For these reasons, several alleged cases of cyber crimes that are directly reported to the Police without preliminary investigations are often susceptible to dismissal, baseless, and unsuitable for litigation purposes after Police have duly reviewed them.

This chapter submits that interrogations of suspects of cyber crime must be in accordance with what the law requires. They should follow the appropriate and proper procedure to litigate offenders. To achieve these milestones, this chapter adopts a log analyzer that extracts and analyzes certain logs of IDSs to explain the

```
[**] [122:1:0] (portscan) TCP Portscan
[**]
[Priority: 3]
08/03-01:59:26.563497 192.168.2.1 ->
192.168.2.2
PROTO:255 TTL:0 TOS:0x0 ID:1176
IpLen:20 DgmLen:159

[**] [125:3:1] (ftp_telnet) FTP command
parameters were too long [**]
[Priority: 1]
08/03-02:03:17.960905 192.168.2.1:64466
-> 192.168.2.2:21
TCP TTL:240 TOS:0x10 ID:0 IpLen:20
```

FIGURE 13.1 Alerts of IDS from a public dataset.

fundamentals of the above issues. Snort in IDS mode is used to sniff traces from the Defcon10 dataset to forensically explore the above issues. The chapter proposes models for classifying suspects and witnesses of cyber crimes that investigators can adopt to critically examine suspects of cyber crimes before reporting them to law enforcement agents. One of the potential contributions of this chapter is its ability to suggest feasible models that investigators may conduct preliminary investigations of suspects and witnesses of cyber crimes in private and corporate environments. The chapter also discusses technical and legal challenges that may influence the evidence of cyber crimes whenever they are to be presented in courts. The chapter further proffers suggestions on how investigators can prepare admissible evidence that may be used to litigate cyber suspects in courts.

13.2 DEFINITIONS OF KEY TERMS

Cyber Physical Systems (CPS): CPS is a concept to describe the security of sensors, processing, storage of large alerts, hardware performance, and software reliability of cyber resources.

Complainant: A person (plaintiff) that institutes legal proceedings against suspects in a court of law.

Cyber crime: A crime that people commit with a mobile device or computer and the Internet.

Cyber criminal: A person that commits an illegal activity with the use of a mobile or computer device and the Internet.

Defendant: A defendant in this context is a person that is being accused or sued of committing a cyber crime. Civil action can be brought against a group of persons, firms, or society.

Intrusion: Intrusion is an infringement into any of the components of the CPS.

IDSs: These are mechanisms for monitoring, detecting, and reporting suspicious activities on components of the CPS

Investigator: A person that examines an allegation of a crime.

Suspects: A person that is being accused of a crime.

Witness: A person that is privy to the details of a crime or conspiracy.

13.3 RELATED TECHNIQUES

The outcomes of interrogations of suspects of cyber crimes may aid their convictions or release. Conventionally, interrogators such as Police often adopt different scientific and theoretical approaches to interrogate suspects (Gehl and Plecas, 2017; Gudjonsson, 1999). For instance, the suggestibility approach has been used by some interrogators to deny the suspects of his/her preferences during interrogations. Accordingly, the suspects may be denied habits like sleep or cigarettes that they have earlier suggested as pleasurable things for them in their leisure time. However, local and international laws do not authorize civilians to adopt them in the course of interrogating suspects of cyber crimes.

Deception or lying has been adopted to interrogate conventional suspects of crimes over the years (Obenberger, 1998). While most laws may not forbid this approach, the suspects might fight back that the interrogators bent to incriminate them by deceiving them to make mistakes and wrong confessional statements. Similarly, an approach that organizes two opposing interrogators to examine suspects have been proposed for studying crimes. In this method, the first interrogator will pretend to be negotiating and supporting the suspect while the second interrogator will pretend and be opposing the suspect (Lewicki and Hiam, 2006). Nevertheless, luring the suspects to confess their degree of involvement in cyber crime also has its shortcomings in courts.

In AMA (2017), certain drugs that can alter the mind of the suspects have been used during interrogations. Nonetheless, these drugs have been openly criticized by medical experts for their capabilities to distort consent and judgment of the suspects. For this reason, the UN has strictly forbidden interrogators from using them. In addition, torture has agelong usage in the interrogation of suspects. However, most information that is gathered from a suspect under torturing is often refutable in courts. Also, they are often linked to confessions interrogators extract from suspects under duress. Above all, the law does not permit civilians to torture or coerce any suspect.

13.3.1 Interrogations of Suspects of Cyber Crimes

Investigations of suspects of cyber crimes are challenging exercises on a daily basis. These procedures often require special preparations in other to achieve better judgment (Gehl and Plecas, 2017; Polivanyuk, 2019). Studies on cyber crimes have shown that some allegations of cyber crimes may be false accusations. Cyber crimes are multifarious activities that include spoofing, an act of hijacking digital transactions in transit, unauthorized probing and scanning of computers, mobile devices, and networks of an organization. Cyber crimes also cover attacks that overload the networks with destructive packets, unauthorized attempt to steal or crack users' passwords and source codes and so on. Thus, ping attacks, ports scanning, and DDoS attacks are cyber crimes.

In Figure 13.2, laws and technicality must converge for interrogators to prepare admissible evidence for courts with the view to litigate cyber-criminals. In cyber laws, the act of closely examining suspects of cyber crimes must not contain or reflect any mistake or misrepresentation of facts.

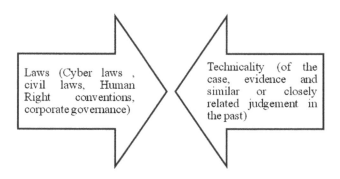

Laws (Cyber laws, civil laws, Human Right conventions, corporate governance)

Technicality (of the case, evidence and similar or closely related judgement in the past)

FIGURE 13.2   Admissible evidence in courts.

The transmission between the interrogators and the suspects should always trigger or generate answering transmissions from the transponders. These should not depict vagueness, contradictory statements, or self-contradictory evidence in the activities of the parties in other to conform to legal requirements.

13.3.2　QUESTIONING SUSPECTS OF CYBER CRIMES

Questioning suspects of cyber crime is a formal request to demand an answer or additional information from the suspects. Unlike the Police that may bring suspects or a witness to a crime into the station for further questioning, civilian investigators must interrogate a witness and suspects under the purview and supremacy of the law of the land. Therefore, such interrogations should cover a set of questions that will closely examine the degree of involvement of the suspects, prior knowledge of suspects about the crime, and the roles each suspect has played in the execution of the cyber crime. Questions should aim to detect inconsistency, improbable, baffling, and self-contradictory evidence from the suspects and witnesses during interrogations of suspects of cyber crimes. Interrogators can adopt the formal technique of systematically questioning the suspects in this context. The investigators may grant the suspects the opportunity to brainstorm and write answers to non-threatening questions related to the incidents (Layton, 2020). This technique is closely related to the statements that Police authorities may demand from suspects in similar scenarios. The suspects cannot be tortured, induced by tear gas, or coerced by any chemical substance (Gehl and Plecas, 2017). Law also forbids civilian investigators from adopting most of the conventional methods that Police may use to interrogate suspects.

Figure 13.3 illustrates four kinds of questions that can be used to interrogate suspects of cyber crimes. The interrogators can further ask four categories of questions from the suspects. They can ask yes/no questions, leading questions, and cross-questions from the suspects. Investigators can ask yes/no or true/false questions to succinctly elicit the response that will be "yes" or "no" or "true" or "false" from the suspects. Similarly, investigators can formulate leading questions during cross-examination of the suspects to clarify impressions the investigators develop about the

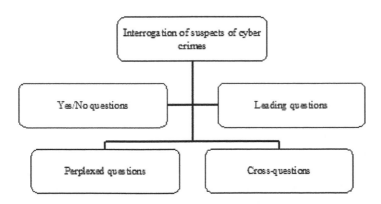

FIGURE 13.3 A model of questions for interrogating suspects of cyber crimes.

answers and body language of the suspects. Thirdly, the investigators may repeat cross-questions for specific suspects on two or more occasions. The questions are varied questions with the aim to judge the truthfulness and establish coherence in the evidence in the interrogations involving many suspects. Perplexing questions are asked to determine facial expressions and psychological emotions of suspects of cyber crime. Some suspects may be unable to think with clarity. Given the fact that some suspects of cyber crime may not possess the mental capacity to commit the crime; therefore, a professional interrogator can adopt this kind of question to specifically determine the cognitive capacity of the suspects.

Moreover, investigators can include the questions they have already asked witnesses to prime suspect(s). It is essential to note that some cyber laws and Fundamental Human Rights, in particular, do not permit investigators to adopt hostile and close questioning of the witnesses to the cyber crimes during cross-examination. Reports of cross-examinations that are done in hostile manners may suffer disparage derogation in courts of law. In addition, it is imperative for investigators to also adopt direct and indirect questioning of the suspects of cyber crimes. The investigators invite only the suspects in the former approach, while both the witnesses and the suspects are invited in the latter approach for cross-examinations. According to Gehl and Plecas (2017), witnesses must be treated with confidential information in the process of using their statements to form reasonable grounds and to charge or acquit the suspects.

13.3.3 THE LEGAL AND TECHNICAL CHALLENGES WITH INTERROGATIONS OF SUSPECTS OF CYBER CRIMES

The suspects and witnesses must not die in the custody of the investigators in the course of their cross-examinations. The basis of law in preliminary interrogations of suspects of cyber crimes is to examine the legality of the entire process. The information the suspects provide may serve as a learning point for the investigators and the accusers, especially if the revelation signifies novelty, serious vulnerabilities, and security lapses in the system. Studies have shown that log analyzers that adopt concepts such as Neural Networks (NNs), Artificial Intelligence (AI), and Genetic Programming (GP) are often criticized whenever they are used for network forensics in realistic settings (Ganapathy et al., 2013).

Thus, in Figure 13.4, this paper presents a log analyzer that adopts clustering of public traces to demonstrate how investigators can extract sources, time of occurrence, descriptions, and destinations of the cyber attacks from IDS logs in network forensics. In other words, Figure 13.4 illustrates how the C++ program can be used to forensically explore intrusive logs and cluster them on the above attributes in other to investigate cyber crimes. The reports of the above log analyzer are demonstrated in Figure 13.5. The results succinctly illustrate the attacks and IP addresses of the computers the intruders have used to compromise the security of computers within simulated networks. A well-documented report will comprehensively itemize various offenses of the suspects. In the case of Figure 13.5, the analysis revealed that the suspects adopted a computer with an IP address of "192.168.2.1" to illegally scan a computer with an IP address of "192.168.2.2". In other words, the perpetrator made a wide and intensely sweeping search of the ports of computers that belong to another

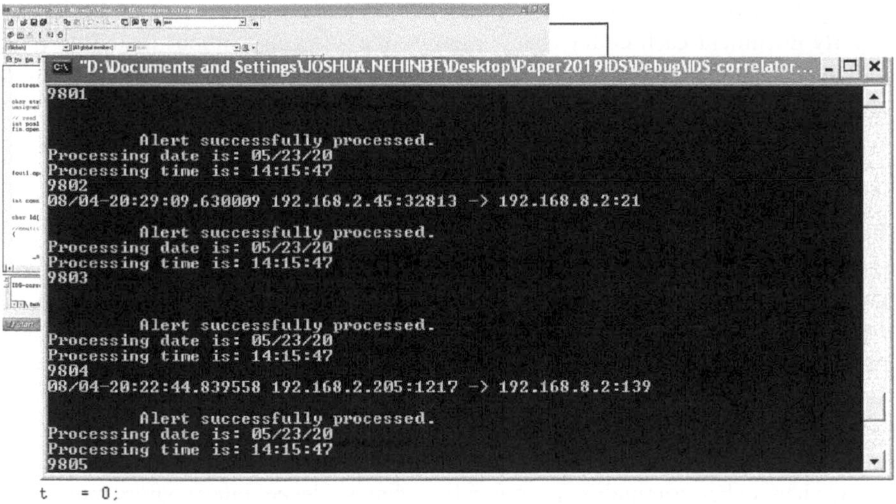

```
9801

            Alert successfully processed.
Processing date is: 05/23/20
Processing time is: 14:15:47
9802
08/04-20:29:09.630009 192.168.2.45:32813 -> 192.168.8.2:21

            Alert successfully processed.
Processing date is: 05/23/20
Processing time is: 14:15:47
9803

            Alert successfully processed.
Processing date is: 05/23/20
Processing time is: 14:15:47
9804
08/04-20:22:44.839558 192.168.2.205:1217 -> 192.168.8.2:139

            Alert successfully processed.
Processing date is: 05/23/20
Processing time is: 14:15:47
9805
```
t   = 0;

FIGURE 13.4 Log analysis of cyber crimes.

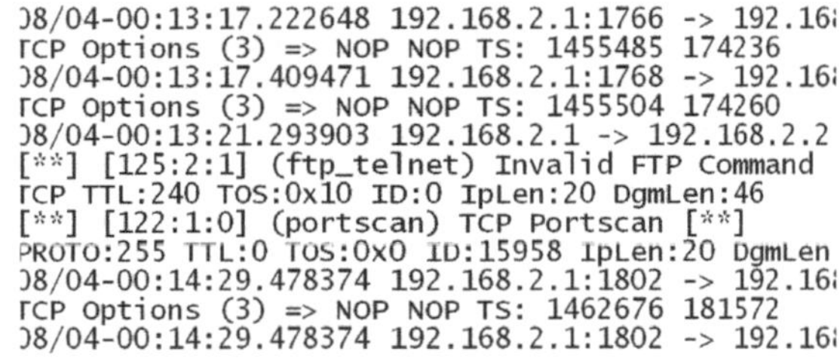

```
)8/04-00:13:17.222648 192.168.2.1:1766 -> 192.16:
rCP Options (3) => NOP NOP TS: 1455485 174236
)8/04-00:13:17.409471 192.168.2.1:1768 -> 192.16:
rCP Options (3) => NOP NOP TS: 1455504 174260
)8/04-00:13:21.293903 192.168.2.1 -> 192.168.2.2
[**] [125:2:1] (ftp_telnet) Invalid FTP Command
rCP TTL:240 TOS:0x10 ID:0 IpLen:20 DgmLen:46
[**] [122:1:0] (portscan) TCP Portscan [**]
PROTO:255 TTL:0 TOS:0x0 ID:15958 IpLen:20 DgmLen
)8/04-00:14:29.478374 192.168.2.1:1802 -> 192.16:
rCP Options (3) => NOP NOP TS: 1462676 181572
)8/04-00:14:29.478374 192.168.2.1:1802 -> 192.16:
```

FIGURE 13.5 Forensic analysis of a public dataset.

person or server with the company without securing approval from the rightful owner and management of the company.

Besides, the analysis demonstrated that the suspects also adopted the same computers to illegally send disruptive packets to disorder, stop and interfere with activities in the networks. The third offense of the suspects was that he/she adopted telnet commands to minutely examine and moved a set of records from one place to another. These offenses may be categorized into three groups. The interrogators must query the suspects to further determine his/her underlying motives for using the commands. It is possible that the suspects used the telnet commands to install Trojans in the system. It is also plausible that the suspects used the commands to extract certain records out of the networks to another location within the networks. Similarly, the commands might have been used to move certain records out of the networks.

System auditors can detect employees by the IP addresses of their computers or mobile systems in organizations. Additionally, system auditors can detect questionable outsiders through collaborations with Networking, regional Registrar, and IP service providers. They can request for full IP blocks of the regions where the suspects reside. In technical terms, system auditors may face different legal challenges while seeking the above information across different international boundaries compare to governments' agencies. In essence, investigators of cyber crimes must properly document each count charge against the suspects without including any ambiguity. It is imperative for them to also permit the suspects to provide his/her responses to each allegation for a fair hearing.

13.4   A MODEL FOR CLASSIFYING SUSPECTS OF CYBER CRIMES

This section presents a new classification of suspects of cyber crimes in Figure 13.6 below to assist investigators of cyber crimes in corporate settings. A financier of cyber crime is a capitalist person that provides unethical financial support to assist a perpetrator of cyber crime. With the new wave of terrorism, interrogators must be prudent in probing the suspects to establish the possibility of cyber crimes that are being outsourced by financiers. In addition, investigators should be aware that suspects of cyber crime may be an abettor. In this case, there is a need to also invite a person that assists or encourages cyber crimes for thorough interrogations. Abettors can also help cyber-criminals carry out cyber crimes. Hence, interrogators must find out the level of involvement of the abettor in the case through series of thought-provoking questions. The culprit is the prime suspect in cases involving cyber crime. The culprit is the person that actually perpetrates the cyber crimes that are being investigated. In law, the culprit or perpetrator of cyber crime is basically not an offender until he/she is being proved by a competent court of law.

Moreover, crime studies suggest that a succorer is a suspect of cyber crime. Investigators need to also interrogate a person that saves or helps a person that is accused of cyber crimes to escape justice or to be temporarily released from custody. Expurgator is another potential group of suspects of cyber crimes. An expurgator can be a technically and non-technically oriented person that helps cyber-criminals to wipe out digitally recorded information that would have helped investigators to track intrusions or trace and arrest the culprits. A person may be charged as a suspect if he/she knowingly or unknowingly edits or removes obscene and other criminal records or makes evidence that is necessary to reveal the intruders invisible to detect.

According to the model in Figure 13.6, a leaker of sensitive information that assists the perpetrator of cyber crime in achieving his/her objective is a potential

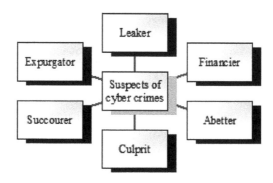

FIGURE 13.6   Categories of suspects of cyber crimes.

suspect that should be interrogated. It is difficult to detect a surreptitious informant that leaks or discloses confidential and unauthorized information to the perpetrator of cyber crimes. The informant may be an employee within the system, or a former staff, a consultant, or someone that is outside the system.

In forensic security, employers must frequently remind their workers that data protection and privacy are issues of official secrets that must be known to a few people in the organization. Therefore, in technical terms, it is unlawful for an unauthorized person to disclose information that is meant to be kept an official secret to a third-party. The scope of the investigations of endemic cyber crimes within a political setting, the site of power, and financial organizations are usually wide than commercial investigations. Improper actions and indecorous behaviors of employees that are tending to or suggestive of moral looseness may call for thorough investigations to establish the relevancy of such unethical actions with the cases under investigation.

13.4.1 A Model for Conducting Preliminary Examinations of Suspects of Cyber Crimes

Pragmatically, we further sought the perspectives of certain professionals on the proposed phases of interrogations of suspects of cyber crimes. Accordingly, 20 professionals from the IT, control, and law/customers' relations departments in commercial settings were contacted to rate the inclusion of the first-line investigators, panel review, and disciplinary committees in the preliminary examinations of suspects of cyber crimes in private and corporate environments.

Figure 13.7 reflects the responses of the above experts in percentages. As shown in Figure 13.8, the respondents collectively agreed that first-line investigators, panel

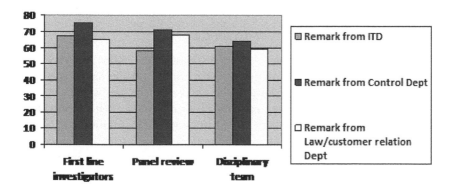

FIGURE 13.7 Phases of reports.

FIGURE 13.8 Preliminary examinations of suspects of cyber crimes.

review, and disciplinary committees are significant stages in the above scenarios. The first-line interrogators must ensure that all logical accesses of the suspects to the networks of the victims (private and organizations) are disabled before they summon the suspects for interrogations. The interrogative team must equally note that it is the requirement in law not to subject the suspects to any form of intimidation or harassment during cross-examinations.

In addition, it is noteworthy for the organizations to also constitute a review board that will independently review the reports generated by the first-line investigators. The review board conducts a panel review of various findings of the previous team. This is the stage whereby the panel must critically review the severity of the allegations and validate the offense with the recommendations suggested by first-line investigators within the available time of reference. The panel may overrule the recommendations at their disposals by acquitting the suspects and close the case if they extensively detect fallacies or insufficient proofs against some or all the suspects in the allegations. Also, the panel may observe that the recommendations of first-line investigators are essentially sensible, and they may decide to wholly or partially uphold them. It is the duty of the review panel to also qualify the cost of punishing the suspects against their inherent benefits to the victims.

The review board (panel review) may subsequently refer the suspects to the disciplinary committee for disciplinary actions if they deem fit. If the case solely involves insiders, the employees may be suspended, fired, redeployed, or fined, depending on their recommendations. The disciplinary committee may be obliged to refer the suspects to the Police for further interrogations if they eventually find out that the organizations have good cases against the suspects and there is minimal risk or no legal implication that is associated with the case. Essentially, all the above teams must discharge their duties independently and in accordance with local laws and UN standards of treating suspects of cyber crimes.

13.4.2 LITIGATION OF SUSPECTS OF CYBER CRIMES

This section discusses salient points that organizations must critically review before they file a formal charge against suspects of cyber crimes. Allegation of cyber crimes that is brought in a court of law against suspects may not necessarily lead to the suing of the accused. Many corporate and private organizations are often expected to win some courts' proceedings against cyber-criminals, but their loss turns out to be a rude shock. Strong suspects may challenge the complainant in court. Some suspects may raise formal objections to protest the ways they were treated during interrogations and to discontinue their proceedings.

In addition, suspects may challenge the illegality and impropriety of a particular line of the allegation or a piece of evidence and claim it is legally wrong. In some cases, the suspects may refute a piece of questioning, a particular witness, and strong facts on other offenses and official matters levied against him/her as improper. Some suspects may argue that they lose credibility and threat to resign their positions. Consequently, suspects may want the case to be continued and would therefore employ the court to rule on all the allegations against them as illegal and impropriety.

A complainant must adequately prepare their evidence to dispute such questioning and statements demanding explanations on the allegations of against their suspects in other not to inadvertently lose the case. Some suspects may plead for an out-of-court settlement. A complainant must be able to preempt the outcome of the proceedings before engaging in a legal battle with suspects of cyber crimes to avoid resource wastage.

13.4.3 MANAGING THE SUSPECTS OF CYBER CRIMES

Cases of cyber crimes must be properly managed to limit unwarranted reactions that competitors may adopt to earn market dominance. The head of interrogators should control all attempts by suspects and team members to walk out during cross-examination of the suspects of cyber crimes. Sympathizers may walk out of the meetings with the suspects without necessarily protesting but deceptively acting as a decoy to frustrate management from continuing with the case. Huge numbers of employees or followers in the society may stage a protest. They may massively troop into the arena of the interrogator's room or premises of the court.

Local and international laws give little protection to witnesses of cyber crimes. Poor management of suspects of cyber crimes may trigger different kinds of industrial actions like a short-term and prolonged strike. Some witnesses may be fired for exposing corporate frauds. Corporate image is at risk whenever workers suddenly stop work in order to press demands for better ways of treating their fellow workers. In some cases, employees may walk out of their departmental meetings or briefing when the issues are raised to demand better procedures of handling cases of cyber crime in the organization.

Investigators of cyber crimes can face a diversity of challenges in different settings. Key members of the company may resign suddenly due to poor handling of the suspects of cyber crimes. The dangers are enormous if significant numbers of employees of the company continue to resign as expressions of anger, protest, and disapproval of their colleagues that were wrongly accused of cyber crimes and unlawfully punished by their management. Negative reactions of employees can cripple the efforts of management to achieve high profitability. These developments may further render corporate governance, operations, and process workflow ineffective. Therefore, investigations of cyber crimes must be thoroughly carried out and handled by knowledgeable people in the corporate organization that is a victim of cyber crime.

13.4.4 DISCHARGING AN INSIDER THAT IS SUSPECTED OF CYBER CRIME

Most organizations do not have procedures for discharging suspects of cyber crime. Witnesses may be discharged on two viewpoints. The allegation may be baseless. In some cases, the allegations may not be profitable for the organizations to pursue to the courts of law. The appointments of some suspects within the organizations may be summarily terminated. The decision-makers in the organization may be liberal with some suspects of cyber crimes by issuing them strong warnings. In this scenario, the suspects must be deployed to different units.

The management must formally write statements to the suspects indicating he/she has been absolved or relinquished of the allegations against him/her. A copy of the document pronouncing that an employee is dismissed or not guilty of criminal charges must be kept in his/her employment file. In some countries like Nigeria, the law requires Banks to also report employees that are caught with criminal activities to the Central Bank of Nigeria (CBN) to deter the employees from working in the banking sector for their lifetime. The dismissed employee may be entitled to certain benefits in the organization. It is the prerogative of the top management of the organization to decide how the benefits of the affected customers will be treated. Cases of cyber crimes may be an opportunity to re-engineer the internal controls of an organization. While restructuring is in progress, the management must approve reforms that are pivoted by the confidentiality, integrity, availability of cyber-physical resources, and protection of civil liberties.

13.5   A MODEL FOR CLASSIFYING WITNESS TO CYBER CRIMES

Information privacy, professionalism, and maturity are required in the process of gathering criminal evidence against suspects of cyber crimes. Furthermore, management must commend a witness that instantly reports cyber crimes early enough so that management of the organization can proactively institute precautionary measures necessary to ward off impending cyber-criminals, dangers, and damages. This chapter proposes five categories of a witness to cyber crime in Figure 13.9.

A potential witness can be a close observer of the suspects while performing cyber crimes. The premise is that a person that looks at suspects while exhibiting some kind of crime may be able to provide useful information to the investigators of cyber crimes. In addition, a witness can be an attestator. An attestator is a person that can demonstrate the genuineness of evidence of cyber crime to authority. Similarly, a witness can be a testifier. A witness in this category is a person that can give evidence of a crime under oath in court. Furthermore, a witness can be an informant. An informant is a person hired or delegated to expose all forms of wrongdoings (like) within an organization in order for top management to proactively stop them. The above model also suggests that a witness can be extracted from the list of security guards in the organization. In this case, a security guard is a person that protects and keeps watch over CPS in an organization.

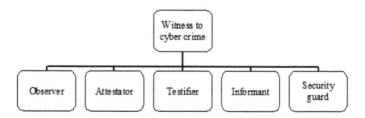

FIGURE 13.9   Witness to cyber crime.

Investigators must be prudent to discern hostile witnesses and his or her motives. It is possible that the witness may not be of a good relationship to the suspects, especially if the witness is an avenger. In this case, his or her testimony may be biased evidence against the suspects. Investigators can use leading questions and cross-examine to detect a hostile witness and his or her motives during facts-findings. On the other hand, it is unlawful and unethical to equate genuine witnesses or treat him/her as a betrayer.

13.5.1 THREATS TO WITNESS TO CYBER CRIME

A witness that notices cyber crime and is courageous enough to report it by stating what happened, where, and how it happened must be safeguarded in order to avoid some unwanted consequences to him/her. Instances whereby witnesses will eventually be punished for revealing confidential information to the interrogators and Police are unwholesome practices that can trigger human rights issues. A witness that is not adequately protected by the organization may be killed by dreadful suspects. Some suspects may send threats or dangerous declarations of his/her intentions and strong determinations to inflict severe harm to a person that has been invited to witness against the suspects. Terrible suspects of cyber crimes may extend threats to exterminate, maim, and bullies to the relatives of witnesses against them. A reprisal attack that is indicative of revenge may be devastated. The technicality of the profession is necessary to curtail all acts of retaliatory actions people can carry out against themselves in the above scenarios.

A witness that receives explicit threats to his/her life may suddenly develop series of mental health disorders. In worst scenarios, threats to the lives of witnesses may induce compulsive behavior, hallucination, restiveness, panic, and paranoid schizophrenia. Another central concern is that a witness that receives reoccurring threats that may even be vague, shadowy, and unpleasant emotions of anticipation of indecipherable misfortunes may require psychiatric solutions to live the rest of his/her life. Some witnesses are victims of cyberwars. They are victims of series of assaults, intimidation, and coercion from cahoots of dangerous suspects through electronic devices and social communication networks. Such victims may develop cyberphobia.

A witness may be implicated to have been selling online contrabands to fictitious customers to criminate, incriminate and punish him/her. A witness that has suffered from excess threats or a victim of being plundering with excessive damages of properties and total destruction of assets may feel the sudden loss of his/her memory and consciousness. Some witnesses to serious courts' cases on cyber crimes have grossly suffered a stroke, stigmatization, colossal ruins, and permanent deformities that have disrupted the normal functioning of their brains. There are instances of restlessness for low-income earners that volunteer to witness if there is need for them to travel long distance. In the long run, self-funding and low-income earners may be incapable of attending numerous adjournments and court proceedings associated with high cases of cyber crimes. Consequently, it is very hard to see people that will voluntarily make themselves available as witnesses to cyber crimes in most settings.

13.5.2 Dismissal of Cyber Lawsuit

Trial judges can formally stop or proceed further with a trial of suspects of cyber crimes in a court of law. The decisions of the judges in this context dismissal often precipitate down to the technicality of the case and the evidence for and against the suspects. For instance, what are the criminal and civil natures of the allegations that are levied against the suspects? Does the materials or exhibits provide against or in favor of the suspects necessary to further adjudicate the case or lawsuit with criminal proceedings or civil proceedings? Hence, suspects of cyber crimes may be absolved of the crimes in a lawsuit if the judges are presented with insufficient, contradictory, and inconsistent evidence.

The goal of most defendants to the suspects of cyber crimes is to obtain a not guilty verdict from the judges of their clients. Criminal and civil technicalities are two legal perspectives to determine the suitability of punishing and deterring serious cyber-criminals as well as the proceeding to further compensate the victims of cyber crime in courts (Lawteacher.net, 2020). In other words, the technicality of Cyber crimes can be construed from a criminal or civil perspective depending on the evidence and exhibit available to the trial judges. The trial judges must establish if the behavior of the defendants or suspects can be construed as an offense against the society, public, or an offense against the state. The judges must further establish if the behavior of the suspects may be constituted as inconsequential judges or as severe injury to the complainant. In a lawsuit, complex cyber crimes may be construed as an offense against the public, society, or government that has also done serious injuries to the representatives of the victims. Therefore, criminal defense lawyers must collaborate with digital security and forensic experts in the organizations to enhance the admissibility of their evidence in the court of law.

The nature of legal protections available to the defendants varies from one country to another. The manner in which complainants have filed the suit, how the judges have decided the case, and standards of valid evidence against the suspects are some of the determinants of dismissals of a lawsuit against suspects of cyber crimes. There should be no disputed issues regarding the material facts in the charges against the suspects of cyber crime. Witness statements must be sworn under oath, and all the recorded questioning sessions on the cyber crime must be transcribed correctly. Otherwise, documentary evidence and sworn witness statements on the case may be regarded as questionable and inconclusive proofs in the court.

Above all, local authorities or councils must individually adopt the legislation compliant with cyber laws for protecting suspects and witnesses to cyber crimes. Compliance with these legislations should be obligatory to federal and local agencies. The act of parading witnesses or suspects of cyber crimes in public spaces, walking them through public spaces, or subjecting them to public ridicule is abusive and unconstitutional. Stripping suspects or witnesses naked, denying them food, disallowing them to have access to their close relatives or friends, and other modes of torturing people are gross violations of the HRBP.

13.6 CONCLUSION

Cyber crimes are economical and political threats to corporate organizations and agencies across the globe. This chapter submits that suspects of cyber crimes must be permitted to voluntarily give information during preliminary interrogations. The reason is that some evidence of cyber crimes may be baffling, conflicting, odd, and self-contradictory after they have been critically evaluated by legal, crime, and forensic professionals. Unfortunately, most of the well-established tactics are not applicable for civilian investigators in practice. Hence, organizations that accuse the suspects of cyber crimes must give the suspects a fair hearing in professional manners before the suspects are reported to the Police. This chapter believes that suspects of cyber crimes must be permitted to clarify from investigators and freely ask questions from their interrogators. Unlike the situation whereby Police would invite the suspects to their stations to possess total control of the situation, preliminary investigators in private and corporate organizations are limited to conference or interrogation rooms. The above have recently generated several debates on the usefulness, legality, and technicality of preliminary investigations of suspects of cyber crimes. Some contenders believe that it is the sole responsibility of the Police to interrogate suspects of cyber crimes.

This chapter has shown that suspects of cyber crimes may be absolved of the crimes due to insufficient, contradictory, and inconsistent evidence. Some suspects may not intentionally commit the crime. It is also plausible that specific malware or Trojans that have compromised and adopted the computer systems of the suspects might be the originators of the accusations the complaints have levied against the suspects. For these reasons, this chapter proposes that preliminary investigations should be carried out by an investigative team of relevant professionals before involving any law enforcement agency. Several issues surrounding this notion have been elucidated above. We argue that accusations that the complaints of cyber crimes presented to the Police may be accepted or rejected by the Police.

Intruders to corporate settings may be charged on many offenses regarding the above issues depending on the depth of understanding of legal and technical terms of the case. Unlawfully intruded into a computer or mobile network is an allegation of uninvited access to the networks without securing approvals from the legitimate owners of the system. Intrusion into corporate networks is a breach of corporate practice. The perpetrator is likely to be guilty or charged for disregard of corporate laws; criminal breach of security rules; failure to agree with terms of contracts; irrecoverably causing a state of devastation and destruction to data files that belong to legally approved organizations, and violations of ethical obligations that are indispensable of best global practice. After due review, Police may decide to dismiss fallacious allegations rather than proceeding to courts of competent jurisdictions that will upturn the case.

Upturning cyber allegations by courts could be disastrous. The suspects may, in turn, sue the complaints about the deformation of characters. As shown above, competitors, journalists, and celebrities may stigmatize firms and private individuals that are famous for lying against suspects. Rivals may misuse wrong accusations as opportunities to traduce, belittle and disparage their competitors. Hence, this chapter recommends that corporate and private individuals should endeavor to adopt models of suspects and witnesses of cyber crimes proposed above. They should always conduct in-depth studies on interrogations of cyber-criminals to ascertain new trends and acceptable standards in the wider society. Local cyber laws should delineate and proscribe online activities that are morally abhorrent. The motives of interrogators in questioning vary. Some questions are meant to elicit curiosity in the suspects or witnesses to a cyber crime. Some questions are asked to confuse or perplex the suspects or witnesses. Questions that make suspects or a witness to be skeptical are done to determine the level of distrust in the evidence the person has earlier given. When you participate in questioning, it is imperative to keep away from asking difficult questions that will make the suspects or the witness confused or bewilder.

13.6.1 Suggestions

Questioning of the suspects and witnesses to cyber crime is a fact-finding mission. Therefore, it is of necessity for the research community to develop a database of questions that civilian interrogators can adopt to determine the level of involvement of suspects in cyber crimes. Presently, the legality of several methods available to law enforcement agencies are restrictive; and hence they are strictly challenged in recent times. For instance, it is unlawful for interrogators to abuse, coerce, beat, injury, or violate the rights of the suspects of cyber crime during their cross-examinations by interrogators.

New cyber laws that will provide a basis for suspects that are victims of torture to seek redress and claim civil damages in courts should be enacted. This chapter further submits that national and international laws that prohibit torture, cruel, inhuman, or degrading suspects of cyber crimes should prescribe strict penalties for the violators of these cyber laws to protect the suspects of cyber crimes. The models in this chapter will be invaluable to the students of computer forensics and forensic specialists working in corporate firms to rapidly know the best performing strategies they can adopt to interrogate suspects and witness to cyber crimes.

Essentially, the death of the suspects of cyber crime might summarily exterminate the case. Therefore, it is very important for the research community to evolve new methods for interrogating suspects of cyber crimes in accordance with legal and United Nations standards.

13.6.2 Future Research

There are numerous ways to extend the research reported in this chapter. For instance, this research work can be extended by exploring the best methods that investigators can adopt to interrogate each of the various categories of the witness and suspects of cyber crime that were stated in this chapter in the nearest future. In addition, there is a need for the research community to develop new models that policemen can adopt to investigate witnesses to suspects of cyber crime in corporate settings. One of the essential strategies to achieve these two milestones is to evolve categorical datasets that can further elucidate the categorization of witnesses and suspects of cyber crimes.

REFERENCES

AMA. "Physician Participation in Interrogation," The American Medical Association, Principles of Medical Ethics: I, III, VII, VIII, 2017. Available at https://www.amaassn. org/deliveringcare/physician-participation-interrogation. Accessed on 27/04/2020.

Center for Strategic and International Studies (CSIS). "Economic Impact of Cybercrime—No Slowing Down," Washington, DC, 2018. Available at: https://www.mcafee.com/ enterprise/en-us/assets/reports/restricted/rp-economic-impact-cybercrime.pdf. Accessed on 27/02/2021.

G. H. Gudjonsson. *"The Psychology of Interrogations, Confessions and Testimony,"* (Wiley Series in Psychology of Crime, Policing and Law), Wiley, Chichester, 1999.

Gehl, R. and Plecas, D. *"Introduction to Criminal Investigation: Processes, Practices and Thinking,"* Justice Institute of British Columbia, 2017.

Ganapathy, S., Kulothungan, K., Muthurajkumar, S., Vijayalakshmi, M., Yogesh, P. and Kannan, A. "Intelligent Feature Selection and Classification Techniques for Intrusion Detection in Networks: A Survey," *Journal on Wireless Communications and Networking,* 2013, no. 271 (2013).

Layton, J. "How Police Interrogation Works: Common Interrogation Techniques," 2020. Available at: https://people.howstuffworks.com/police-interrogation1.htm. Accessed on 28/04/2020.

Lewicki, R.J. and Hiam, A. *"Negotiation: A Working Guide to Making Deals and Resolving Conflicts,"* Jossey-Brass, San Francisco, CA, 2006.

Nehinbe, J.O. "Methods for Reducing Workload During Investigations of Intrusion Logs," PhD Dissertation, University of Essex, UK, 2011.

Obenberger, J.D. "The Law and the Skin Trade in the Windy City: Police Deception," 1998. Available at: http://my.execpc.com/~xxxlaw/GP03-98.htm. Accessed on 24/04/2020.

Perri, P.S., and Lichtenwald, T.G. "Exposing Fraud Detection Homicide," 2018. Available at: https://www.all-about-psychology.com/support-files/exposing_fraud_detection_ homicide.pdf. Accessed on 26/08/2021.

Lawteacher.net. "The Admissibility of a Judgment in a Criminal Proceeding Vis-a-vis," 2020. Available at: https://www.lawteacher.net/free-law-essays/criminal-law/admissibility-of- a-judgment-in-a-criminal-proceeding-law-essay.php?vref=1. Accessed on 08/03/2021.

Polivanyuk, V. "Interrogation of Suspects in Investigating Computer Crime," Computer Crime Research Center, 2019. Available at: http://www.crime-research.org/library/ Polivan1003eng.html. Accessed on 26/04/2020.

14 Fault Analysis Techniques in Lightweight Ciphers for IoT Devices

Priyanka Joshi and Bodhisatwa Mazumdar
Indian Institute of Technology Indore, Indore, India

CONTENTS

DOI: 10.1201/9781003140023-14

14.1   SECURITY IN IoT ENVIRONMENTS

With growing trends in Cyber-physical systems and the IoT, tiny, embedded devices are rapidly proving to be ubiquitous, whereas IoT services are exhibiting pervasive characteristics. With a huge impact on the business environment and social life, IoT systems are becoming the fastest growing technology. Technology is gradually permeating all aspects of human life, for instance, education, healthcare and business, and financial transactions. The development of design methodologies of such systems must ensure that the IoT systems are safe, secure, and reliable. On the one hand, a myriad of systems connected to computer systems creates a large attack surface for the adversaries to observe and manipulate the data in transit and storage. On the other hand, the consequences of many conventional security attacks, such as DoS, have magnified the safety concerns of the systems. In many cases, safety and security are tightly entangled. For instance, in May 2015, the flight control systems of the Boeing 737 operated by United Airlines were claimed to have been while the airplane was still in flight [1]. The stakes of security and safety of such IoT systems are very high, given the criticality of operations. In general, IoT devices embedded in industrial control systems or smart city networks are projected to have a much longer lifespan than present-day consumer electronic devices. The enormous diffusion of computation components in industrial control systems IoT systems has created a large demand for robust security with the ever-growing demand for millions of interconnected devices and services.

The devices in IoT environments produce, process, and transfer large volumes of confidential data that even comprise privacy-sensitive information. Consequently, they form appealing targets to the attackers [2–7]. For ensuring the correctness and safety of operations, it is imperative to incorporate the security of the underlying devices, especially data and the code, against unauthorized modifications. In the recent past, many security vulnerabilities have been identified in such embedded devices [8–12]. This has created new challenges in the implementation of secure IoT systems exhibiting several functionalities and security features at a minimal cost.

Cloud computing has become an integral part of IoT systems wherein the devices store data for processing and subsequently transfer it to other devices after authorization. It provides a model to access a shared pool of configurable computing resources on demand. It is envisioned that users and organizations with different capabilities can store and process their data in data centers that may be located in geographical regions that are far away from the user. With services that are primarily driven by data-intensive computations, security and data privacy have become major challenges. With the advent of the widespread use of IoT devices, cyber attacks are more of a physical threat rather than remotely operated. With massively present IoT devices that are connected to the cloud, malicious adversaries can find opportunities to physically control these devices and gain access to the user data that need to essentially remain private. Over the period of last decade, embedded devices are experiencing ever-growing complexity and have become an integral part of IoT systems. In such a scenario, vendors are struggling to incorporate security into the designs that were earlier emphasized for efficiency and performance.

IoT systems have started incorporating cryptographic protocols in their embedded devices to alleviate safety and security concerns. IoT devices have to operate in highly resource-constrained environments with limited computing and storage capabilities, design and implementation of cryptographic modules have led to the emergence of *lightweight cryptography*. In this branch of cryptography, emphasis is laid on the development of highly resource-efficient cryptographic primitives and, therefore, tailor-made for embedded platforms. Although these cryptographic primitives are mathematically proven to be secure, multiple implementation-based attacks have emerged that have threatened cryptographic protocols of global standards and adopted across the world. This chapter presents a discussion on FA attacks, a variant of implementation-based attacks, on lightweight ciphers.

With the advent of IoT systems, the user community has witnessed the widespread adoption of small computing devices, such as smart cards, sensor nodes, radio frequency identification (RFID) tags, etc. Consequently, the requirement to address many security and privacy issues for such small devices has rendered the task of applying conventional cryptographic standards quite challenging. In such standards, the tradeoffs between resource requirements, security, and efficiency, that have been widely accepted for desktop and servers are infeasible for resource-constrained devices. For instance, many IoT devices use low-power microcontrollers to afford only a small fraction of security features and computing power. Moreover, conventional cryptographic encryption and authentication primitives may incur either a large latency or a large power consumption that is unacceptable for devices in IoT environments.

14.2   LIGHTWEIGHT CIPHERS FOR IoT SYSTEMS

In typical sensor networks deployed in IoT systems, a large number of sensors are connected to a central hub. All these sensors are battery-powered or solar-powered. Cryptographic primitives pertaining to encryption and authentication are used in communication channels between the central hub and sensor nodes to provide secure computation and communication, authenticity, and message integrity. However, due to limited energy availability, security requirements are always overhead in the actual device functionality. Similarly, RFID tags are deployed for the identification of objects such as devices, animals, and sometimes even human beings. With the emerging success of RFID technology, it is increasingly used in supply chain and consumer electronic applications for the protection of services by cryptographically secure RFID tags. This has led to the requirement of hardware-efficient cryptographic primitives, for example, hash functions, authentication protocols, and symmetric encryption schemes.

Along with the device functionality and cryptographic requirements, the circuit in such tags is limited to not more than 4000 gate equivalents. The corresponding implementation of cryptographic primitives and authentication protocols must satisfy the tight timing requirements of ISO 18000-63 standard. Moreover, such devices should not consume more than 30–50 μW of energy in addition to no energy consumption peaks as it would severely limit the range of devices. In addition, the authentication protocols must have minimal round-trip communications. Further, the number of transmitted bits should be minimized with a small probability of false authentication. Over the past decade, various attack scenarios have emerged on RFID systems, such as impersonation and relay attacks, tag cloning or unauthorized copying, unauthorized tracking and monitoring, information leakage, privacy infringements, and DoS attacks [13].

In the last two decades, the field of cryptographic research has made significant progress in the design and development of lightweight cryptographic primitives, such as PRESENT, KATAN/KLATAN, SIMON and SPECK, RC5, TEA, XTEA, LED, and hash functions, such as SPONGENT, PHOTON, and QUARK. The proposals of cipher primitives and hash functions are targeted toward optimizing metrics, such as memory requirements, area footprint, and speed or throughput of the primitive. These performance metrics are proposed while considering whether the implementation platforms are hardware or software.

14.3 DESIGN CONSTRAINTS FOR HARDWARE IMPLEMENTATIONS

If the cryptographic primitive for encryption and authentication is targeted for hardware implementation, memory requirement, and size of implementation in a combination called gate equivalent, which quantifies the size of the circuit that implements the cryptographic primitive. The throughput of the implementation is reported in bytes/second, which relates to the bits of the plaintext message processed within a time unit. The performance metric depends on the characteristics of the circuit under consideration, i.e., a maximum clock frequency of circuit operation or the gate count of the circuit. Memory is usually the costliest component of the lightweight cipher's implementation targeted for hardware platforms. For instance, in many cases, implementations need to store the full internal state or the round key and subsequently executed one round over n clock cycles.

14.4 DESIGN CONSTRAINTS FOR SOFTWARE IMPLEMENTATIONS

The typical platform for software implementation of cryptographic primitives primarily comprises the microcontrollers. The relevant metrics in software implementations include the RAM access, the size of code and data, and the performance of the implementation that is calculated in bytes per CPU clock cycle. A comparison of these metrics across a spectrum of crypto-primitive algorithms and different implementations of the same algorithms is performed by *Fair Evaluation of Lightweight Cryptographic Schemes (FELICS)* [14]. This framework considers implementing a block cipher or a stream cipher as the input. The framework produces the respective RAM consumption, code size, and execution time of the crypto-primitive as the output. In the specified figure-of-merit (FoM), the block size and key size parameters are mentioned in bits. The code size and RAM consumption parameters are computed in bytes, and the execution time is measured in the CPU clock cycles required to complete a task. However, all these performance parameters are inter-linked. For instance, loading operation from RAM into CPU registers and its inverse operation is a costly operation in terms of power consumption. Hence, restricting the count of such load and store operations results in a reduction in both RAM access and execution time.

The implementations have shorter internal states and smaller key sizes to satisfy resource constraints of lightweight crypto-primitives in IoT devices. Several light-weight primitives use 64-bit blocks; the smaller block size heads to reduced memory consumption in software and hardware. Furthermore, this also implies that the algorithm is devised for shorter-length messages.

In lightweight block ciphers, many crypto-primitives, such as Sboxes, are proposed as involution mappings that decrease the cost of implementing the decryption module. Moreover, appropriate Feistel structures can also be made to utilize a single implementation for both encryption and decryption processes. The International Organization for Standards and the International Electrotechnical Commission (ISO/IEC) standard 29192, which deals with lightweight cryptography, states that the minimum-security strength for any crypto-primitive is 80-bit security. However, the standard recommends that a system that requires security for more prolonged periods should adopt a primitive with at least 112-bit security strength.

Another crypto-primitive where dedicated lightweight algorithms are required is hashing operations. Whereas conventional hashing functions require large memory to store the internal state and block of data that they are operating on, such requirements degrade the performance in the case of resource-constrained platforms. This creates the need for dedicated lightweight hash functions.

14.5   COMMUNICATION PROTOCOLS

The communication protocols that form the basis of the interconnected IoT devices comprise an encryption form, also need to be lightweight. For example, cell phone communication is based on GSM, and 3G networks, whose specifications mention that the communication must be encrypted using either of the following protocols: 3G, KASUMI, SNOW, ZUC, A5/1(/2/3).

Bluetooth and its lower energy consumption version, Bluetooth Smart connects devices which are in physical proximity with each other. In this standard, E0 stream cipher was proposed in the initial specification, but it was subsequently replaced with Advanced Encryption Standard (AES). Present-day Wi-Fi connections are protected by WPA, WPA2, and WPA3 security protocols. To interconnect wireless IoT devices, IEEE 802.15.4 standard is used, which applies the Zigbee protocol.

14.6   CRYPTO-PRIMITIVES WITH LIGHTWEIGHT DESIGN

With the evolution of lightweight block ciphers in the past decade, some design characteristics have emerged that comprise the choice of nonlinear operations and key schedules. The nonlinearity property is imparted by Sboxes in SPN block ciphers and Feistel ciphers. Typical lightweight Sbox implementations comprise lookup tables (LUTs) and bitsliced implementations. For considering the nonlinearity of arithmetic operations, only modular addition operation is considered, which is exhibited by the Addition-Rotation-XOR (ARX) class of block ciphers and stream ciphers.

LUT-based implementation of Sboxes offers an approximately optimal set of cryptographic properties with a single operation. Nevertheless, as they involve storing an entire set of output values and frequently accessing those values, it incurs a significantly high cost. Moreover, accessing LUT incurs leaking some information of the secret key. Hence, Sboxes with smaller dimensions, such as 4x4 Sboxes, are used in Piccolo [15], PRESENT [16], and SKINNY [17].

Sboxes can also undergo bitsliced implementation, which avoids the usage of table lookup, thus minimizing the resource count even further. In such an implementation, AND and XOR bitwise operations are executed on words of w bits, therefore allowing Sbox execution in parallel for multiple instances. A bitsliced implementation is related to a small area for hardware implementation, which implies that such implementations will perform efficiently in hardware. In software platforms, besides a limited number of logical operations, bitsliced implementations provide simple security arguments based on a wide-trail strategy. Hence, bit-slices Sboxes are the preferred choice in the design of lightweight algorithms. There exist many block ciphers, for example, ASCON [18], iScream [19], PRIDE [20], KETJE, RECTANGLE, Noekeon, RoadRunneR, Fantomas, and FLY, which adopt such bitsliced Sboxes in their architecture.

14.7   ADDITION-ROTATION-XOR (ARX)-BASED CIPHERS

ARX-based block ciphers are based on modular addition operations to impart nonlinearity to the ciphers, whereas word-wise rotations and XOR provide diffusion property to such ciphers. Due to carrying propagation, the modular addition operation generates highly nonlinear most significant bits at the output. SPARX is an instance of ARX-based provably secure cipher toward differential and linear attacks [21]. In hardware implementations, modular addition is expensive, which increases significantly with larger word lengths. For software implementations, modular addition is relatively cheaper. Hence, a number of ARX-based ciphers give the best performance in microcontroller platforms, for example, Chacha20, Salsa20, LEA, SPARX, SPECK, XTEA. The lightweight ciphers can be divided into two categories: ultralightweight cryptography and the second is IoT cryptography. We consider two instances, one in each of these classes:

- KTANTAN performs encryption of blocks comprising at most 64 bits with 256 rounds and a master key of 80 bits.
- LEA encrypts messages of 128-bit block size by employing XOR and rotation operations along with 32-bit modular additions, with key dimensions of 128, 192, and 256 bits.

The circuit required to implement nonlinear Sbox in KTANTAN consumes very few gates, whereas LEA has among the best performance metrics in FELICS. KTANTAN is hardware-oriented, whereas LEA is software-oriented.

14.8   ULTRALIGHTWEIGHT CRYPTOGRAPHY

This class of ciphers deals with the most constrained use cases. For instance, RFID tags that are used in access control based on challenge-response do not require sizable 256-bit security. Since the throughput of such devices is not high, attacks requiring a large number of queries or data are rendered infeasible for such devices. Such cryptographic algorithms run on very low-cost devices that are usually not connected to the Internet and can be easily replaced on account of limited shelf life. Applications of these algorithms comprise RFID tags, RAIN tags, remote car keys, smart cards, and memory encryption. Multiple instances of block ciphers and stream ciphers, such as Grain, Trivium, SKINNY, PRESENT, PRINCE, PHOTON, and KTANTAN, fit this requirement. For such ultralightweight algorithms, the cipher can be either a stream cipher or a block cipher, the key size must be at least 80 bits, and the block size must be at least 64 bits. As such devices have very limited computing power, the volume of output ciphertexts shall be very small. While considering a maximum query size of 250 plaintext/ciphertext pairs and retaining k bits of security of the embedded key, it reduces the upper bound on the total number of rounds in the cipher.

14.9   IoT CRYPTOGRAPHY

On the other hand, an IoT cryptographic primitive usually runs on a battery-operated device with low computing power connected to the Internet. Since several such devices operate in environments where an adversary may have physical access, such as security cameras placed in an open environment, it is essential that SCA countermeasures should be implemented easily. Furthermore, as IoT devices carry out various tasks, multi-purpose microcontrollers are used for executing cryptographic operations rather than digital electronics circuits. Therefore, efficient software is essential in this case. For IoT systems, many lightweight cryptographic protocols have been devised targeting software implementations rather than hardware implementation, including SPARX, FLY, Pride, Chaskey, LEA. In implementations of every algorithm, the key will occupy registers with an intended platform as a microcontroller. The block size is a minimum of 96 bits, whereas the key size must be 128 bits. With a larger block size, the internal state and the key shall fit in the registers of a typical microcontroller in IoT systems.

14.10   KEY SCHEDULE OPERATION

The class of lightweight ciphers differs most from the conventional ciphers in terms of key schedule operation. Encryption functions running on standard computers can have a complex key schedule, wherein a master key is used to generate the round keys. A round key is produced and stored in memory before the round function starts executing. On the other hand, lightweight ciphers are heavily resource-constrained; hence, the incurred area in RAM size or gate area is unacceptable.

14.11   FA ATTACKS

The idea of FA attacks was proposed in a seminal paper by Boneh et al. [22]. In FA, the adversary mounts a malicious perturbation during the execution of a crypto-graphic algorithm. The ease of fault injection and afterward retrieving the key depends on the nature of fault, spatial and temporal fault characteristics, and the fault propagation properties of the underlying cipher. The classes of fault characteristics comprise bit-flip, byte fault, stuck-at faults, or random faults. The spatial and tempo-ral fault characteristics refer to the register location and round where a fault is injected. The fault propagation properties in a cipher refer to the round operations that diffuse the fault across the round and the fault injection timing. We can classify the FA techniques based on the principle of the mounted attacks into the categories discussed below.

14.11.1   DFA

Faults are injected with constant spatial and temporal fault characteristics in the DFA technique of FA [23–25]. Followed by required fault injection, the adversary com-putes the differentials between pairs of correct and faulty ciphertexts to obtain the secret key. Over the past two decades, DFA has been used on many block ciphers that are in present-day use. The authors in [26] illustrated that only a single fault injection is required to recover the entire 128-bit key of the AES block cipher. SPN-based block ciphers have been widely analyzed against DFA with many successful attack attempts. The schematic for generic SPN-based block ciphers is shown in Figure 14.1.

The block cipher includes n rounds, with a block size of N bytes. Moreover, every round comprises a nonlinear substitution layer of Sboxes, a linear diffusion layer, a key mixing function that adds the round key RKi with the intermediate state. The permutation layer, also called the diffusion layer, involves a permutation of its inputs that can be in the order of bits, nibbles, or bytes, followed by a linear transformation. This transformation is usually implemented with a maximum distance separable (MDS) matrix. Both the Sbox operation and the diffusion function operate on the bytes of the intermediate state.

In this case, if an input byte of a diffusion layer is modified due to the fault of the input bit-flip, the differential change propagates to multiple output bytes of the diffu-sion layer in that round. This propagation depends on the diffusion layer's differen-tial branch number, which defines the minimum number of modified bytes output of the diffusion layer for one affected byte at the input. These bytes are called faulty bytes in FA, whereas in differential cryptanalysis, these bytes are called active bytes. For the diffusion layer, the differential branch number value gives the lower bound on the number of active bytes at the output of the diffusion layer for a given number of its faulty input bytes. Suppose an attacker injects a single-byte fault in the inter-mediate state before the $(n-1)^{th}$ round (second last round/penultimate round). Hence, the respective differential value is α ($\neq 0$). An input byte fault affects δ bytes if the diffusion layer's differential branch number is δ. In this case, a byte fault propagates to $(\delta-1)$ bytes ($\alpha_{d,0}, \alpha_{d,1}, \ldots, \alpha_{d,\delta-2}$) at the output of the diffusion layer, where d refers

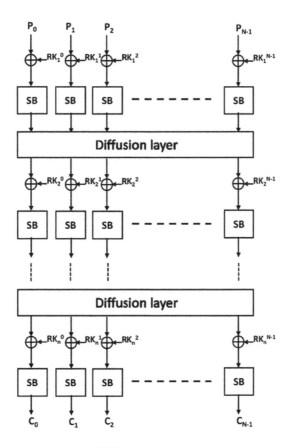

FIGURE 14.1 General structure of SPN block ciphers.

to the mapping of the diffusion layer. All such fault bytes undergo nonlinear transformation by the substitution layer. The FA adversary can provide the following representation of the output faulty bytes of the diffusion layer in terms of faulty and fault-free ciphertext bytes (C_i, C_i^*).

$$\alpha_{d,i} = S^{-1}\left(C_{d,i} \oplus K_{d,i}^{n-1}\right) \oplus S^{-1}\left(C_{d,i}^* \oplus K_{d,i}^{n-1}\right)$$

Where $\delta \in \{0, \ldots, \delta - 1\}$ and S^{-1} the inverse Sox operation. As the adversary knows the difference of the inverse Sbox input values, $C_{d,i} \oplus C_{d,i}^*$, and also as he knows the difference distribution table of Sbox, he can reduce the entropy of the last round key $K_{d,i}^{n-1}$ by eliminating the key candidates that satisfy the preceding equation. The FA adversary can adopt a divide-and-conquer approach by repeating the same steps of the attack to retrieve more bytes of the last round key of the cipher. In this analysis, the differential of the nonlinear component, i.e., the Sbox assists the key retrieval process.

14.11.2 FAULT SENSITIVITY ANALYSIS (FSA)

In [27], Li et al. proposed this mode of attack. FSA identifies the intermediate data-dependent transition bits by analyzing the pairs of correct and faulty ciphertexts. Fault sensitivity is a condition wherein whenever a fault occurs, and certain information is always available to the adversaries. For instance, whenever the fault injection intensity increases, attackers can pinpoint the conditions at which faults start occurring or the fault becomes stable. Hence, if the correlation between the fault sensitivity and the cipher's intermediate state that is a function of the key is known, then the FSA attack can easily obtain the secret key embedded in the IoT device.

14.11.3 DIFFERENTIAL FAULT INTENSITY ANALYSIS (DFIA)

The class of differential behavior analysis (DBA) attacks and safe error attacks (SEA) arise from the condition that whether a fault occurred during encryption indeed results in an incorrect computation [28, 29]. This analysis depends on the fault propagation characteristics in the implementation that is dependent on the secret key. In other words, it uses the fact that some sections of the key determine whether a fault injection results in a faulty ciphertext. This can also be termed as fault behavior analysis (FBA), wherein the attacker only observes the behavior of the device behavior under perturbation and does not need access to the ciphertexts. However, they use fault models like stuck-at fault that may be expensive to accomplish [30].

14.11.4 SEA AND DBA

These attacks exploit and combine principles of FA with side-channel analysis techniques, such as differential power analysis (DPA), for key recovery [31, 32]. The attack requires access to fault ciphertexts only. It selects key hypotheses related to a non-uniformly distributed faulty intermediate state. Owing to statistical characteristics, DFIA requires a large number of queries with faulty ciphertexts. Despite such requirements, a moderately equipped adversary can use DFIA to break the security of block ciphers.

14.12 FAULT INJECTION METHODOLOGIES: SEMI-INVASIVE AND NON-INVASIVE METHODS

Over the past two decades, security researchers have shown different methods by which faults can be injected in smartcard ICs possessing cryptographic protocols. This section discusses some semi-invasive and non-invasive methods to physically inject faults in cryptographic devices. These methods have been effective on numerous security systems. However, for widespread threat identification, conventional fault injection methods must refrain from costly actions and operations, such as depackaging an IC device to access the inside electrical contacts. Some of the conventional fault injection methods are as follows.

14.12.1   POWER SURGE

A power spike or power surge is a sudden fluctuation in the power supply voltage to an FPGA device running an IoT subsystem or a smartcard IC, that is beyond tolerance limits. A sudden spike in voltage levels leads to faulty computations or faulty data load operation in registers, and the fault characteristic is a function of the transition shape, spike voltage value, and spike timing of the voltage signal. Each of these characteristics can be controlled by the attacker to produce a spike that induces the intended fault during the execution of the cryptographic primitive.

14.12.2   CLOCK GLITCH

Attackers, in glitch attacks, generate a sudden variation in the clock pulse supply to the circuit implemented on the targeted device such as smartcard IC; this disturbed clock pulse is called a clock glitch. The fault gets induced when the instantaneous pulse of the clock glitch exceeds the operational clock frequency to produce the correct ciphertext output. The produced clock glitches cause skip instructions during program execution on software platforms and setup time breaches in hardware implementations. Clock glitches are one of the cheapest and popular modes of fault injection techniques on ICs comprising cryptographic primitives and hardware [28, 33–35].

14.12.3   LASER INJECTION

Laser injections have been used to inject faults in non-volatile memory or electrically erasable programmable read-only memory (EEPROM) of smartcard ICs [36–38]. Subsequently, highly focused laser beams were used to pinpoint specific faults in targeted register sections of a microcontroller. For instance, the attacker can target the instruction register of the microcontroller and, with the help of laser beam injection, toggle the opcode bits to induce an instruction skip fault with precise timing and repeatability. This attack is of semi-invasive nature, which at times may require depackaging the IC to inject the laser beam.

14.12.4   ELECTROMAGNETIC INJECTION

Electromagnetic radiations can be used to induce bit-set/reset faults in data bits in the shift registers in FPGA or in the memory of the microcontroller. The process induces eddy currents in registers by placing a coil conducting alternating current near the target register. As a result, transient faults are generated at bit level in the register locations. This is a non-invasive attack that can work through a smartcard wherein the power of the electromagnetic pulse can be enhanced by an amplifier [39–41].

14.13   TYPES OF FAULTS

The injected faults can be categorized into two classes based on the time duration an injected fault stays in the system: permanent or transient. Permanent faults are created when the IoT device with cryptographic primitive is subjected to extreme operational conditions, such as high temperature, high voltage, or exposure to focused laser beams for a long time period. Such exposure can result in register bits stuck to zero value (lose their ability to switch to one value) or stuck to one value (lose their ability to switch to zero value). Such stuck-at fault conditions can force certain bits in an intermediate state to get permanently assigned 0 or 1 values. This may lead to a loss in entropy of the secret key that can be exploited by an attacker.

Transient faults lead to sudden non-persistent flips in the bit registers of the target device due to sudden glitches in the power supply or clock line or short pulses from the laser gun. Both under-powering the device or clock glitches lead to setup time violations in the hardware, hence injection of errors [42, 43]. These faults stay momentarily and do not damage the device. As the injected fault is temporary, it affects only one function call to encryption or decryption function.

14.14   FAULT MODELS

In FA attacks, there exist different fault models in the attack implementations. A class of attacks considers random single-bit faults, which require a high precision fault injection setup to flip exactly one bit in the intermediate state register. However, if injected, such faults lead to a large reduction in key entropy in a given number of queries. In addition, single-byte faults belong to a more generic class of attacks and present an acceptable tradeoff between the practicality of injection and the time required to perform the analysis. This renders the attack practically exploitable. With increasing ease of injecting faults, multiple byte fault models require a low-cost fault setup. Although it results in complex FA, its ease of mounting makes it a potential threat to cryptographic implementations.

14.15   COUNTERMEASURES TO MITIGATE FA ATTACKS

In the preceding section, we have provided different variants of FA techniques, mechanisms to inject faults, and the corresponding fault models that exist for IoT-based lightweight cryptographic implementations in software and hardware. In totality, FA attacks form a serious threat to IoT products which house such cryptographic implementations and hence require sound countermeasure to thwart such threats.

The initial mechanisms to thwart such attacks were based on the detection of an occurrence of faults in any execution of an implemented security algorithm. This solely relies on the characteristic of the fault that most faults that are injected are stealthy and transient in characteristics. Such fault detection mechanisms comprise concurrent error detection (CED) mechanisms, which rely on fault identification using a specialized circuit that exhibit one or multiple forms of temporal, spatial, or data redundancy. For instance, an easy method can be to execute a redundant copy of the computation instructions of the algorithm at fault-sensitive points for each instance of algorithm execution. The outputs of these redundant computations are then compared to catch a fault. When a fault is detected, the countermeasure mechanism either halts the execution or infects the ciphertext by randomizing the ciphertext to render it unintelligible for information extraction of the last round key. Some popular redundancy techniques comprise spatial redundancy, such as hardware datapath redundancy or partial spatial redundancy.

Temporal redundancy techniques are based on repeating each operation twice. However, this reduces the throughput of the circuit by 50%. This reduction is resolved by using double-data-rate (DDR) techniques, wherein non-adjacent stages in pipelined implementation are operated by clock signals with opposite polarities. Consequently, the clock cycles required are reduced to half of the initial count, with a little cost to the maximum clock frequency. With improved technology scaling, the feasibility of having efficient implementations is still a major challenge for such DDR techniques. Alternatively, sliding-based pipelined implementation of block ciphers has been proposed that aims toward saving the number of clock cycles. In this method, separate pipeline stages execute primary and duplicate computations, thus providing a sliding feature (Figure 14.2).

In a way, timing redundancy and spatial redundancy are very similar to each other as the primary and extra computations are performed parallelly. Nevertheless, as the computations are done in separate stages of the pipeline, every stage can use its dedicated hardware for the computations; the supplementary hardware is required only for multiplexers and comparators require. In countermeasure implementation of

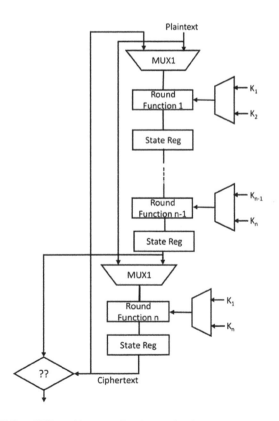

FIGURE 14.2   Sliding CED architecture for time redundancy.

AES-128 design, with just a hardware overhead of 2.3%, the sliding approach retains 90% of the throughput compared to unprotected implementation.

In addition to timing redundancy-based countermeasure, information redundancy-based on code-based detection techniques exist, which incur even lesser overhead. In this technique, a set of checksum bits are produced at each computation stage based on an appropriate error detection code. The checksum bits get diffused by the encryption algorithm along with the original message. The checksum bits can be used to ensure the correctness of the produced ciphertext. Some of these techniques comprise robust codes and parity-based redundancy architectures. Moreover, hybrid redundancy techniques were proposed as a combination of multiple classical redundancy techniques described in this section. An instance of such a countermeasure technique can be a combination to render a computation and its inverse operation. That can be used to compare the input of each computation with the output of its respective inverse operation. The redundancy can be at the algorithmic level, round function level, or atomic operation level.

14.16   CONCLUSION

This chapter presents the essence of data security for IoT devices, which gains importance due to their strictly resource-constraint specifications associated with limited processing power and memory storage. To impart such security measures, the design and implementation of cryptographic modules have led to the emergence of *lightweight cryptography*. First, we present different variants of encryption algorithms proposed to address the design concerns in lightweight cryptography. We present different types of FA techniques, such as DFA, DFIA (DFA), DFIA (DFA), FSA, followed by different types of fault injection mechanisms and fault models that a generic FA adversary employs to recover knowledge about the secret key embedded in the device. Last but not least, we focus on different countermeasures that have been proposed to thwart FA, which depend on the fault injection model and power of the fault analysis adversary.

REFERENCES

1. Perez, E.: "FBI: Hacker claimed to have taken over flight's engine controls," *CNN*, 2015. [Online]. Available: https://www.cnn.com/2015/05/17/us/fbi-hacker-flight-computer-systems/index.html. [Accessed: 24-Jun-2021].
2. Neumann, P.: The RISKS digest. *Committee on Computers and Public Policy*, 22(87), 2021. Available: https://en-academic.com/dic.nsf/enwiki/749328 [Accessed: 17-Sep-2021].
3. Poulsen, K.: Slammer worm crashed Ohio nuke plant network, 2003. Available: https://www.theregister.com/2003/08/20/slammer_worm_crashed_ohio_nuke/ [Accessed: 17-Sep-2021].
4. Levy, E.: Crossover: Online pests plaguing the offline world. *IEEE Security Privacy*, 1(6), 2003. Available: https://rampages.us/keckjw/wp-content/uploads/sites/2169/2014/11/Myths-and-Facts-for-Control-System-Cyber-security.pdf [Accessed: 17-Sep-2021].
5. Byres, E., and Lowe, J.: *The myths and facts behind cyber security risks for industrial control systems*. PA Consulting Group, London, 2004.

6. Vijayan, J.: Stuxnet renews power grid security concerns. Computerworld, Jul. 26, 2010. [Online]. Available: http://www.computerworld.com/s/article/9179689/Stuxnet_renews_power_grid_security_concerns. [Accessed: 26-Aug-2021].

7. Kabay, M.: Attacks on power systems: Hackers, malware, 2010.

8. Costin, A., Zaddach, J., Francillon, A., and Balzarotti, D.: *A large-scale analysis of the security of embedded firmwares.* In *USENIX Conference on Security Symposium.* USENIX Association, 2014. Available: https://www.usenix.org/node/184450 [Accessed: 17-Sep-2021].

9. Cui, A., and Stolfo, S. J.: *A quantitative analysis of the insecurity of embedded network devices: Results of a wide-area scan.* In *Annual Computer Security Applications Conference (ACSAC).* ACM, New York, 2010.

10. Nicol, D. M.: Hacking the lights out. *Scientific American*, 305, 70–75, 2011.

11. Koscher, K., Czeskis, A., Roesner, F., Patel, S., Kohno, T., Checkoway, S., McCoy, D., Kantor, B., Anderson, D., Shacham, H., and Savage, S.: *Experimental security analysis of a modern automobile.* In *IEEE Symposium on Security and Privacy (S&P).* IEEE, New York, 2010.

12. Checkoway, S., McCoy, D., Kantor, B., Anderson, D., Shacham, H., Savage, S., Koscher, K., Czeskis, A., Roesner, F., and Kohno, T.: *Comprehensive experimental analyses of automotive attack surfaces.* In *USENIX Conference on Security.* USENIX Association, 2011.

13. Bhardwaj, A., Subrahmanyam, G. V. B., Avasthi, V., and Sastry, H.: Security algorithms for cloud computing. *Procedia Computer Science*, 85, 535–542, 2016.

14. Dinu, D. -D., Biryukov, A., Großschädl, J., Khovratovich, D., Le Corre, Y., and Perrin, L.: *FELICS—fair evaluation of lightweight cryptographic systems.* In *NIST Workshop on Lightweight Cryptography 2015.* National Institute of Standards and Technology (NIST), 2015. Available: https://csrc.nist.gov/csrc/media/events/lightweight-cryptography-workshop-2015/documents/papers/session7-dinu-paper.pdf [Accessed: 17-Sep-2021].

15. Shibutani, K., Isobe, T., Hiwatari, H., Mitsuda, A., Akishita, T., and Shirai, T.: Piccolo: an ultra-lightweight blockcipher. In *International Workshop on Cryptographic Hardware and Embedded Systems*, pp. 342–357. Springer, Berlin, 2011.

16. Bogdanov, A., Knudsen, L.R., Leander, G., Paar, C., Poschmann, A., Robshaw, M.J.B., Seurin, Y., and Vikkelsoe, C.: PRESENT: An ultra-lightweight block cipher. In *International Workshop on Cryptographic Hardware and Embedded Systems*, pp. 450–466. Springer, Berlin, 2007.

17. Beierle, C., Jean, J., Kölbl, S., Leander, G., Moradi, A., Peyrin, T., Yu, S., Sasdrich, P., and Sim, SM: The SKINNY family of block ciphers and its low-latency variant MANTIS. In *Annual International Cryptology Conference*, pp. 123–153. Springer, Berlin, 2016.

18. Dobraunig, C., Eichlseder, M., Mendel, F., and Schläffer, M.: Ascon v1. 2. Submission to the CAESAR Competition, 2016. [Online]. Available: https://competitions.cr.yp.to/round3/asconv12.pdf. [Accessed: 26-Aug-2021].

19. Grosso, V., Leurent, G., Standaert, F. X., Varici, K., Durvaux, F., Gaspar, L., and Kerckhof, S.: SCREAM & iSCREAM side-channel resistant authenticated encryption with masking. Submission to CAESAR, 2014. [Online]. Available: https://competitions.cr.yp.to/round1/screamv1.pdf. [Accessed: 26-Aug-2021].

20. Albrecht, M. R., Driessen, B., Kavun, E. B., Leander, G., Paar, C., and Yalçın, T.: Block ciphers–focus on the linear layer (feat. PRIDE). In *Annual Cryptology Conference*, pp. 57–76. Springer, Berlin, 2014.

21. Dinu, D., Perrin, L., Udovenko, A., Velichkov, V., Großschädl, J., and Biryukov, A.: Design strategies for ARX with provable bounds: Sparx and LAX. In Cheon, J. H. and Takagi, T. (eds.) *Advances in Cryptology—ASIACRYPT 2016, Part I, Lecture Notes in Computer Science, 10031*, pp. 484–513. Springer, Heidelberg, 2016.

22. Boneh, D., Millo, R., and Lipton, R.: On the importance of checking cryptographic protocols for faults. In *Advances in Cryptology EUROCRYPT97*, pp. 37–51. Springer, Berlin, 1997.

23. Kim, H.: Differential fault analysis against AES-192 and AES-256 with minimal faults. In *2010 Workshop on Fault Diagnosis and Tolerance in Cryptography (FDTC)*, pp. 3–9. IEEE, New York, 2010.

24. Mukhopadhyay, D.: An improved fault based attack of the advanced encryption standard. In Preneel, B. (ed.) *Progress in Cryptology—AFRICACRYPT 2009. Lecture Notes in Computer Science*, 5580, pp. 421–434. Springer, Berlin, 2009.

25. Piret, G., and Quisquater, J.J.: A differential fault attack technique against SPN structures, with Application to the AES and KHAZAD. In *Cryptographic Hardware and Embedded Systems, CHES 2003: 5th International Workshop*, pp. 77–88. Springer, Berlin, 2003.

26. Tunstall, M., Mukhopadhyay, D., and Ali, S.: Differential fault analysis of the advanced encryption standard using a single fault. In *Information security theory and practice. security and privacy of mobile devices in wireless communication*, pp. 224–233. Springer, Berlin, 2011.

27. Li, Y., Sakiyama, K., Gomisawa, S., Fukunaga, T., Takahashi, J., and Ohta, K.: Fault sensitivity analysis. In *Cryptographic Hardware and Embedded Systems, CHES 2010, 12th International Workshop*, Santa Barbara, CA, pp. 320–334, 2010. Available: https://www.iacr.org/archive/ches2010/62250310/62250310.pdf [Accessed: 17-Sep-2021].

28. Blömer, J., and Seifert, J.P.: Fault based cryptanalysis of the advanced encryption standard (AES). In Wright, R.N. (ed.) *Financial Cryptography. Lecture Notes in Computer Science*, 2742, pp. 162–181. Springer, Berlin, 2003.

29. Robisson, B., and Manet, P.: Differential behavioral analysis. In *Cryptographic Hardware and Embedded Systems - CHES 2007, 9th International Workshop*, Vienna, Austria, pp. 413–426, 2007.

30. Li, Y., Hayashi, Y. I., Matsubara, A., Homma, N., Aoki, T., Ohta, K., and Sakiyama, K.: Yet another fault-based leakage in non-uniform faulty ciphertexts. In *Foundations and Practice of Security*, pp. 272–287. Springer, Berlin, 2014.

31. Fuhr, T., Jaulmes, E., Lomne, V., and Thillard, A.: Fault attacks on AES with faulty ciphertexts only. In *2013 Workshop on Fault Diagnosis and Tolerance in Cryptography (FDTC)*, pp. 108–118. IEEE, New York, 2013.

32. Ghalaty, N., Yuce, B., Taha, M., and Schaumont, P.: Differential fault intensity analysis. In *Workshop on Fault Diagnosis and Tolerance in Cryptography (FDTC)*. IEEE, New York, 2014.

33. Anderson, R., and Kuhn, M.: Tamper resistance: a cautionary note. In *Proceedings of the Second USENIX Workshop on Electronic Commerce*, 2, pp. 1–11. USENIX Association, 1996.

34. Anderson, R., and Kuhn, M.: Low cost attacks on tamper resistant devices. In *International Workshop on Security Protocols*, pp. 125–136. Springer, Berlin, 1997.

35. Kömmerling, O., and Kuhn, M.G.: Design principles for tamper-resistant smartcard processors. *Smartcard*, 99, 9–20, 1999.

36. Biham, E., and Shamir, A.: Differential fault analysis of secret key cryptosystems. In Burton, S.K. (ed.) *Advances in Cryptology—CRYPTO 1997. Lecture Notes in Computer Science*, 1294, pp. 513–525. Springer, Berlin, 1997.

37. Maher, D.P.: Fault induction attacks, tamper resistance, and hostile reverse engineering in perspective. In *International Conference on Financial Cryptography*, pp. 109–121. Springer, Berlin, 1997.

38. Naccache, D., and M'Raihi, D.: Cryptographic smart cards. *IEEE Micro* 16(3), 14–24 (1996).
39. Alberto, D., Maistri, P., and Leveugle, R.: Investigation of electromagnetic fault injection effects on embedded cryptosystems. In *First Workshop on Trustworthy Manufacturing and Utilization of Secure Devices, TRUDEVICE 2013*, 2013.
40. Dehbaoui, A., Dutertre, J.M., Robisson, B., and Tria, A.: Electromagnetic transient faults injection on a hardware and software implementations of AES. In *2012 Workshop on Fault Diagnosis and Tolerance in Cryptography (FDTC)*, pp. 7–15. IEEE, New York, 2012.
41. Moro, N., Dehbaoui, A., Heydemann, K., Robisson, B., and Encrenaz, E.: Electromagnetic fault injection: towards a fault model on a 32-bit microcontroller. In *2013 Workshop on Fault Diagnosis and Tolerance in Cryptography (FDTC)*, pp. 77–88. IEEE, New York, 2013.
42. Selmane, N., Guilley, S., and Danger, J. L.: Practical setup time violation attacks on AES. In *EDCC*, pp. 91–96. IEEE Computer Society, New York, 2008.
43. Barenghi, A., Bertoni, G., Parrinello, E., and Pelosi, G.: Low voltage fault attacks on the RSA cryptosystem. In *FDTC*, pp. 23–31, 2009.

Index

Page numbers in *Italics* indicate figures.

Milton Keynes UK
Ingram Content Group UK Ltd.
UKHW031532071024
449327UK00005B/110

9 780367 686505